THE TRANSPORTATION OF SOVIET ENERGY RESOURCES

THE TRANSPORTATION OF SOVIET ENERGY RESOURCES

Matthew J. Sagers
Adjunct Professor of Geography, Weber State College
Center for International Research, Bureau of the Census

Milford B. Green
Associate Professor of Geography,
University of Western Ontario

ROWMAN & LITTLEFIELD
Publishers

ROWMAN & LITTLEFIELD

Published in the United States of America in 1986
by Rowman & Littlefield, Publishers
(a division of Littlefield, Adams & Company)
81 Adams Drive, Totowa, New Jersey 07512

Library of Congress Cataloging-in-Publication Data

Sagers, Matthew J.
The transportation of Soviet energy resources.

Bibliography: p. 155
Includes indexes.
1. Power resources—Soviet Union—Transportation.
2. Energy industries—Soviet Union—Location.
I. Green, Milford B. II. Title.
HD9502.S652S24 1986 380.5′24 86-15508
ISBN 0-8476-7504-1

88 87 86
5 4 3 2 1

Printed in the United States of America

Contents

Figures

Tables

Preface

Soviet energy resources are the largest in the world, yet energy problems have not bypassed the USSR. A major aspect of current Soviet energy difficulties lies in the problem of transporting energy. This is because of the USSR's large size (one-sixth of the world's land surface) and the spatial disparity between energy resources and demand. A key dichotomy exists between the industrialized and densely settled "European" portion of the USSR and the thinly populated, but energy-rich, eastern portion (Siberia). Most of the demand for energy originates in the European portion, as it contains 75-80 percent of the Soviet population, industry, and social infrastructure. In contrast, the Siberian portion contains only about 10 percent of the Soviet population, but almost 90 percent of the country's energy resources.

This monograph analyzes the transportation of Soviet energy resources. The purpose is to determine the general pattern of movement for each of the main forms of energy (gas, crude petroleum, refined products, coal, and electricity), to identify constraints in the transportation system that inhibit efficient flows, and to evaluate the prospects for future developments, based upon this analysis of the system.

To accomplish this, each energy-transportation system is modeled as an abstract, capacitated network consisting of supply nodes (production sites), demand nodes (consumption sites), and bounded arcs (transport linkages). The essential characteristics of each are supply availabilities (production), demand levels (consumption), and transport capacity, respectively. The optimal flows and associated costs for each network are determined by applying the out-of-kilter algorithm, a network allocation model, to each abstracted system. This is done for each network using 1980, the most recent year for which reasonably complete statistics are available, as the base year.

Generally, the models provide justification for the post-1980 developments in each system. Often these developments appear as solutions to the bottlenecks and transport problems identified in the analysis. Thus overall, the current Soviet energy program seems to be a rational response to existing problems. The major findings and speculations of the study are summarized below for each of the major forms of energy.

Natural gas has been given the leading role in the Soviet energy program, with a 48 percent increase in output between 1980 and 1985, with a 30 percent increase planned for 1986-1990. Its relatively recent development means the gas system has changed considerably in recent years, so gas flows are modeled for 1970, 1975, and 1985 in addition to 1980. The modeled flows closely replicate the actual flows, implying that the Soviet gas system is operating near optimum efficiency. Our models of the Soviet gas network also indicate that it becomes successively more congested between 1970 and 1980, consonant with the maturation of the Soviet gas industry.

The major constraint in the pipeline system in 1980 was in the main lines from West Siberia, with other major bottlenecks along the Northern Lights route, at the main distribution points of Ostrogozhsk and Novopskov (located south of Moscow), and in the lines to the Transcaucasus. Subsequent pipeline developments in the network have alleviated these constraints, indicating that their construction was warranted.

A scenario for 1985, incorporating five major new lines (six were actually constructed, although the last was not fully operational in 1985), while increasing supply and demand at several key nodes, shows an improvement in the overall efficiency of the distribution network, with the most serious constraints in the system being alleviated. This confirms the view that the major pipeline construction program of the 11th Five-Year Plan was necessary. Also the rather significant increase in gas exports contemplated between 1980 and 1985, including those to Western Europe, does not particularly tax the gas system with the additional lines installed.

In the transportation of crude petroleum, the hypothetical flow patterns generated by the model in 1980 generally agree with what little is known about actual flows, implying that the petroleum network also is operating efficiently. The major constraints in the system in 1980 were in the European USSR, particularly in the older oil-producing regions in the Caucasus, rather than in the east at the origin of most of the USSR's petroleum. This problem has been alleviated to a considerable extent by the construction of additional pipelines in the region since 1980, so again the model correctly anticipates subsequent developments. In particular, the new Surgut-Novopolotsk pipeline greatly increases the efficiency of petroleum transmission. The model shows the pipelines from West Siberia to the European USSR to be operating close to capacity in 1980. This demonstrates why a new pipeline from West Siberia to the European USSR (between Kholmogory and Klin) was constructed in 1984-1985, to accommodate the growth in petroleum production in West Siberia during the 11th Five-Year Plan (1981-1985).

In contrast, the movement of refined products is relatively inefficient. About twice as many ton-kilometers actually are required in distributing refined products in the USSR as are generated in the optimal flow model. This discrepancy is primarily due to the mismatch which exists among the various products between local consumption needs and refinery output mix, giving rise to a variety of cross-hauls between regions. This implies that Soviet transportation planning is adequate for homogeneous commodities in unique transportation systems (e.g., gas, crude petroleum, electricity), as their distribution is relatively efficient, but apparently does not perform very well with the complexity of heterogeneous products on common carriers.

A second simulation of refined product flows, incorporating the refinery expansion of the 11th Five-Year Plan, indicates that the locations chosen are generally justified. Beyond this, the model indicates that the most likely areas for future refinery construction are the western Ukraine and the Far East.

The simulation of Soviet coal flows explains and justifies post-1980 plans for the sector, mainly in the development of a coal-by-wire approach for the lower quality coal of some eastern coal basins. This is because the model generates a massive movement of coal from the east at volumes that must be straining railroad facilities to the limit. The model also indicates that further expansion of production at Ekibastuz appears warranted. However, expansion is not possible without more transport capacity. This dilemma appears to be resolved in the Soviet plans to expand production at Ekibastuz, using the fuel to power mine-mouth generating stations, and transmitting the energy to the Urals and central Russia with extra-high-voltage electrical lines. An alternative solution to the transport problem that also is supported by the results of the model is the conversion of the Central Siberian rail line into a specialized coal carrier to transport the higher quality Kuznetsk coal to the European USSR.

Our model of the Soviet electrical system indicates that it is operating under far more strain than it was in 1975. A higher percentage of arcs are capacitated, and the problem proves to be infeasible without adjusting for greater dispersion of consumption and including some planned extra-high-voltage (EHV) lines as if they were operational. The main problem in the Soviet electrical network is the lack of transmission capacity, particularly in the Urals region. As a result, the model demonstrates the definite need for the planned EHV network of 1150 KVA lines in the Siberia-Kazakhstan-Urals region. These arcs are very heavily utilized in the model. In contrast, a simulation incorporating the planned 1500 KVA (DC) line from Ekibastuz to central Russia indicates that the project does not appear justified because of very limited utilization and perverse flow patterns on some of the added arcs.

Acknowledgments

The completion of any book is the result of the collaboration of many friends and colleagues of the authors. This being the case it seems only proper to acknowledge their efforts if only in this small way.

The work presented here would not have been possible without the financial support provided by the National Council for Soviet and East European Research. In addition we would like to thank Weber State College for financial support. For this support we are very grateful. Not all the support we received was in the form of funds but rather took the form of the provision of skilled labor.

The preparation of the manuscript was possible in no small way through the efforts of several members of the Department of Geography, at The University of Western Ontario. In particular, Diane Shillington, who labored long and hard over the typing, Gordon Shields and Patricia Chalk of the cartographic section who produced the figures, and Val Noronha who produced the computer drawn base maps.

This book which has been trying at times owes a considerable debt for its completion to the support provided by our wives Donna Green and Janice Sagers who kept us sane.

Finally we would like to acknowledge the Center for International Research, Bureau of the Census for allowing Matt the opportunity to complete this project.

And finally, of course, any errors, omissions, or opinions are solely the responsibility of the authors.

We would also like to thank Tim Heleniak and Jeff Nylund for checking the figures and references.

THE TRANSPORTATION OF SOVIET ENERGY RESOURCES

I
Introduction

A. BACKGROUND FOR THE STUDY

Soviet energy resources are the largest in the world, but energy problems have not bypassed the USSR. Factors contributing to the Soviet Union's energy difficulties include escalating domestic demand, rising production costs, long-term export commitments to Eastern Europe, opportunities for export to the West, and a deteriorating reserve position for petroleum (Office of Technology Assessment, 1981; Dienes and Shabad, 1979; Stern, 1981). A key concern in these difficulties is the ability of the USSR to provide sufficient energy for domestic and export demand. Given the USSR's vast resource base, any failure to produce (and deliver) the desired quantities of energy cannot be attributed to inadequate resources (Stern, 1981, p.4; Office of Technology Assessment, 1981). Instead, other factors, particularly geographical aspects, become fundamental.

The Soviet energy economy operates in a geographic space which comprises almost one-sixth of the earth's land surface (22,402 square kilometers of territory), and also includes the Eastern European countries of CMEA (Council for Mutual Economic Assistance). The large areal extent of this system makes spatial aspects such as circulation and movement of considerable importance. The spatial dimension is highlighted by the key dichotomy existing between the more industrialized and densely settled "European" portion of the USSR and the thinly populated, but energy-rich, eastern portion (Siberia). Most energy demand originates in the European portion, as it contains about 75-80 percent of the Soviet population, industry, and social infrastructure (Mints, 1976; Rumer and Sternheimer, 1982). In contrast, the Siberian portion contains only about 10 percent of the Soviet population, but almost 90 percent of the country's energy resources (Mints, 1976; Dienes and Shabad, 1979). These two macro-regions are separated by thousands of kilometers.

Therefore, the growing energy demand of the European USSR, as well as exports to Eastern Europe and the West, will have to be met with net energy transfers from Siberia. This poses an enormous challenge for the various transportation systems involved, as they must bear the burden for resolving the USSR's spatial imbalance in energy supply. Styrikovich and Chernyavskiy (1979, p.11) predict that by the end of the century, the European USSR's share of total Soviet energy demand will still be 65-70 percent, requiring a flow of energy from the eastern portion 2.5 times that of 1975, which would amount to about 900 million tons of standard fuel. Campbell (1983, p.209) suggests a more reasonable estimate for the year 2000 is a flow of 1,117 million tons of standard fuel, or about 38 percent of a likely estimate of total production for that year (p.206). In contrast, about 360 million tons of standard fuel were involved in the East-West flow in 1975 (Campbell, 1983, p.209), or 23 percent of national production (<u>Narkhoz SSSR 1982</u>, p.143). The Soviets recently reported that 700 million tons of standard fuel, or 35.7 per cent of national production, was transmitted from the Eastern region to the European USSR in 1980. It was also estimated that the flow in 1985 would be 950 million tons or 42.2 per cent of national production (<u>Vestnik akademii nauk SSSR</u>, no. 12, 1985).

The economy's shift to the less accessible Siberian resources has been one of the major factors responsible for rising production costs and some of the other problems being experienced by the Soviet energy industries (Tretyakova, 1985). This shift has made the problem of transporting energy a matter of primary importance, as in some cases (e.g., natural gas and coal), development and production have been constrained by the availability of transport.

In the past, the Soviet transport sector generally has responded well to the heavy demands placed upon it despite inadequate support. Traditionally, the Soviet government has been extremely reluctant to invest in expanding transport capacity; transport services have been provided only when and to the degree absolutely necessary (Hunter, 1968; Hardt _et al._, 1966). At the same time, heavy industrial output has expanded rapidly. This combination has kept demand constantly pressing against the available supply of transport services. This has been particularly true in the case of the railroads, the mode which has traditionally borne most of the USSR's transport burden. Incredible demands have been placed upon the railroads, and for the most part they have responded impressively. For example, rail freight traffic increased 6 fold between 1950 and 1984 (_Narkhoz SSSR 1984_, p.338) without commensurate increases in capital and labor inputs. Recently however, the railroads have begun to falter from the increasing strain (Hunter and Kaple, 1982; 1983).

A growing imbalance between supply and demand is threatening to make transport the key constraint restricing economic growth, just as it was in the 1930s (Hunter and Kaple, 1983, p.i). An important element in the escalating transport burden in the Soviet economy has been the recent changes in fuel supplies and the evolution of regional fuel mixes (Dienes, 1983a, p.405). Essentially, production of fuels in the European portion of the USSR, where most demand is located, has declined as its energy resources have become depleted. Thus the needs of the economy are being increasingly met by eastern fuels which must be transported long distances, often thousands of kilometers. The heavy transport burden this imposes is taxing the economy. The large capital resources and operating costs required to move these fuels play a significant role in the poor performance, and still poorer prospects, for the Soviet economy (Dienes, 1983a, p.405; Schroeder, 1985, pp.61-63).

The availability of energy in the USSR, and transport's role in assuring this, are important. The Soviet economy, like those of the other centrally planned economies in CMEA, has a relatively high energy intensiveness, showing a high correlation between growth in energy inputs and the annual rate of economic growth, with no evidence of any recent decline (Office of Technology Assessment, 1981; Dienes, 1983a; Hewett, 1984, pp.102-104). This reflects the traditional Soviet strategy of "input infusion". Growth in output has been obtained through increased quantities of inputs (particularly labor and raw materials) (Kirichenko, 1981; Pyzhkov, 1982; Greenslade, 1976).

This phase of "extensive" industrialization is now supposedly giving way to a new strategy of "intensive" development consonant with the period of developed socialism (Mints, 1976; Alisov, 1976). This change is largely due to growing relative scarcities of these inputs as Soviet labor reserves have dwindled (Feshbach and Rapawy, 1976) and the stocks of accessible, high quality raw materials have been exploited. These changes have meant more stress upon improvements in factor productivity and more efficient use of resources as growth stimulants. This has required increased imports of Western technology to help increase productivity in the Soviet economy, and thus concomitantly, increased exports of natural resources (particularly energy) to pay for the technology (Dohan, 1979).

Therefore there are a variety of consequences if enough energy to fulfill Soviet domestic needs as well as export requirements to both the Eastern European CMEA countries and the West cannot be delivered. These include an adverse impact upon the aggregate rate of growth of the USSR's economy, both from a shortage of energy inputs and from needed Western technology, a shortage of hard currency to purchase Western technology and grain, the possibility of greater social unrest in Eastern Europe and altered economic and political relations within the Soviet bloc if the long-standing Soviet commitment to fulfill the energy requirements (particularly in oil and gas) of Eastern Europe cannot be met (see Hewett, 1984).

B. NATURE OF THE INVESTIGATION

The purpose of this monograph is to analyze a specific aspect of the spatial dimension of the Soviet energy system, that of efficiently transporting energy in its major forms (oil, coal, natural gas, and electricity) from sites of production to those of consumption. The study will specifically focus upon the general pattern of flow within each energy-transportation system and upon identifying major impediments to efficient flows and other transport problems.

This is accomplished by modeling each energy-transportation system as it existed in 1980 as an abstract capacitated network. The year 1980 is utilized because it is the most recent year for which enough information is availble to support the modeling effort. In each case, this network consists of supply nodes (production sites), demand nodes (consumption sites), and bounded arcs (transport linkages). The essential characteristics of each of these elements are supply availabilities (production), demand levels (consumption), and transport capacity, respectively. The optimal flows and associated costs for each network are determined by applying a network allocation model, the out-of-kilter algorithm (Chapter II), to each abstracted system. This pattern of optimal flows is compared with what little is known of actual flows. From this, general observations on the relative efficiency of the system can be made. This also provides some insight on the wisdom of certain planned or current projects or developments, as the analysis is able to identify the need for some or the dubiousness of others.

Each of the following chapters discusses one of the forms of energy. The allocation of natural gas is presented in Chapter III, while that for crude petroleum is given in Chapter IV. The distribution of refined petroleum products is the subject of Chapter V. Coal flows are discussed in Chapter VI and electricity in Chapter VII. Chapter VIII represents a summary of the major findings for each system and an evaluation of their relative efficiencies and prospects.

II
The Out-of-Kilter Algorithm

The technique used in the following chapters in the analysis of the various energy distribution systems in the Soviet Union is the out-of-kilter algorithm. The explanation that follows is a pedagogical one for those readers who are not interested in the mathematical base of the algorithm or the mechanics of its operation. For the mathematically inclined reader, an excellent exposition of the algorithm may be found in Minieka (1978).

Assume that the following problem is faced by a transportation or energy ministry. Given the railroad network shown in Figure 2.1; what is the most efficient distribution of coal such that all power plants receive adequate coal supplies and no rail line is overburdened. Efficiency is defined as the delivery of coal such that the total transportation cost of delivery is minimized.

To describe the solution process for this problem, a number of terms require definition. The coal fields, power stations and junction of the rail lines are all nodes. The coal fields are supply nodes, the power stations are demand nodes, and the junctions are intermediate nodes. The nodes are an abstract representation of the locations of facilities on the surface of the earth. The nodes are represented by small roman letters such as x.

The rail lines connecting the nodes together can be thought of as allowing the movement or flow of coal in only one direction. In this example, the allowed direction is from west to east. These directed lines are called arcs, to be denoted by ordered pairs of roman letters such as (x,y). x denotes the tail of the directed link and y denotes the head of the arrow. These arcs may represent any sort of connection of nodes such as pipeline or

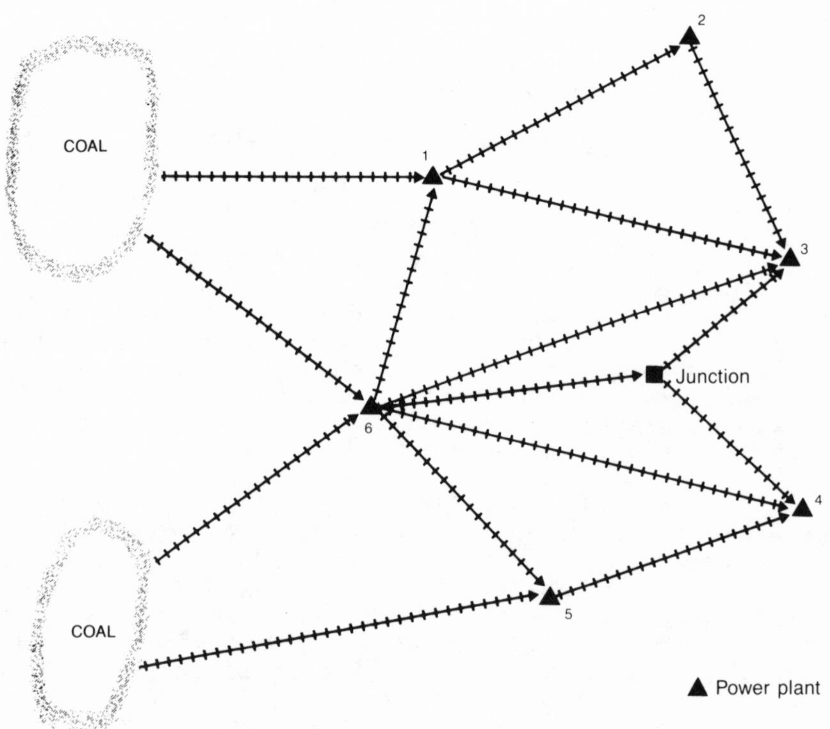

Fig. 2.1 Example Network

electrical power transmission lines. A collection of nodes and arcs is called a network.

The amount of coal moved in the distribution network is called the flow or flow units of coal. A flow in a network is a way of sending objects from one node to another by traveling along the arcs in the allowed directions. Flows shall be labeled as f(x,y) or $f_{x\ y}$.

If it is assumed that each rail line has a maximum amount of coal that can be shipped across it in a period of time, the rail line or arc has an upper capacity denoted as u(x,y). Typically the upper capacity or upper bound of an arc is physically determined, such as in the case of a diameter of a pipeline.

An arc may also have a lower non-negative limit associated with it. That is, a flow of some minimum amount must travel along the arc. This might occur when an arc must be utilized with a flow at least equal to this lower bound in order to be operated economically. A minimum amount may be determined by the physical characteristics of the network such as the level of petroleum necessary to maintain pressure in a pipeline. This lower bound of an arc is denoted l(x,y).

The shipment of coal along an arc in our example has a cost associated with it. This cost may be expressed as the per unit transport cost or the distance between the nodes that define the arc. The analyses in this research utilize distance as the cost utilizing an arc. This is justifiable on the grounds that there generally exists a linear or nearly linear relationship between charged transportation cost and distance. Distance is more easily measured and is readily calculated; hence its use.

Any solution to a problem of distribution must meet a set of constraints. The constraints of our problem are that any flow along an arc must be equal to or greater than the lower limit l(x,y) of the arc as well as be less than or equal to the upper limit u(x,y) of the arc. In addition, the flow may only go in the one specified direction. There are additional constraints that must be met for a mathematical solution to be achieved, but these do not affect interpretation and are hence not discussed here.

The solution procedure is as follows:

1) A set of flows along the arcs is arbitrarily selected. These may be zero. These flows need not initially meet the upper and lower arc limit requirements.

2) Determine the kilter numbers. Kilter numbers indicate whether an arc should have increased flow allocated to it, have flow removed from it, or should be left unaltered. The kilter number denotes the amount by which an arc is out-of-kilter, or the amount it is above the upper limit or below the lower limit of the arc. Each arc has a kilter number. The basic idea of the procedure is to reduce the kilter numbers of all the arcs to zero while not increasing the kilter number of any arc. Kilter numbers are only a means to help achieve a solution and have no meaning outside of the procedure.

3) Once it is determined that an arc has a kilter number greater than zero, the algorithm proceeds to modify the distribution of flows to drive the kilter number to zero.

4) The procedure continues to change the flows' distribution until either all kilter numbers of all arcs are zero, or until no further arcs can be put in kilter. If all kilter numbers are zero, an optimal solution has been found. If the procedure terminates without putting all kilter numbers to zero, no optimal solution is possible and the problem is said to be infeasible. If the problem is infeasible, modifications must be made to the upper or lower limits of some of the arcs.

The infeasibility of a problem often arises when an arc or set of arcs can not carry sufficient flow to allow arcs elsewhere in a flow sequence to be made feasible. That is, to ensure the flow along an arc is between or equal to its upper and lower bounds. Arcs that prevent a solution are called bottlenecks because they constrain flows.

How does the out-of-kilter algorithm reduce the kilter number of an arc say (x,y) to zero? It is first determined whether an increase or a decrease in the flow of arc (x,y) is needed. If an increase is needed, the algorithm searches for a sequence of arcs from a coal field to that arc that can carry to necessary increase without violating the upper and lower limits on any of the arcs in that sequence. If such a sequence of arcs can not be found, the algorithm makes changes in the node prices by a specified amount.

The node prices constitute an artificial price system. In our example, the node price at a node x denotes the delivered price of coal at that node say p(x). These prices are determined only through the use of an arbitrary base price at the mine site and arc costs. The node price structure reflects what the delivered price of coal would be at each node if the distribution of flows under consideration is taken as a solution. The node prices do not necessarily reflect actual delivered prices of coal since factors other than transport cost can affect price.

Given that such a node price system is created, it can be used to determine which arc flows should be increased and which should be decreased. This is done through the creation of opportunity costs. Each arc has an opportunity cost. The opportunity cost of an arc is defined to be

$$\overline{c_{ij}} = c(x,y) + p(x) - p(y)$$

where

$\overline{c_{ij}}$	is the opportunity cost of arc (x,y)
p(x)	is the node price at node x
p(y)	is the node price at node y
c(x,y)	is the arc cost of arc (x,y)

Opportunity costs may be negative, positive or zero.

For a feasible flow on arc (x,y) to be considered optimal it must be true that:

if $\overline{c_{ij}} < 0$ then $f(x,y) = u(x,y)$

if $\overline{c_{ij}} > 0$ then $f(x,y) = l(x,y)$

if $\overline{c_{ij}} = 0$ then $l(x,y) \leq f(x,y) \leq u(x,y)$

If the opportunity cost of arc (x,y) is less than zero, the flow f(x,y) must be the maximum possible. If the opportunity cost is greater than zero, then flow f(x,y) should be minimized. If the arc has its lower bound l(x,y) set to zero, then the flow f(x,y) must also be zero. Opportunity costs of zero indicate that in the optimal solution, flows in these arcs would not affect the solution.

In the optimal solution, the existence of a negative opportunity cost implies that additional savings could be gained if the upper bounds u(x,y) of an arc (x,y) were increased. Opportunity costs of positive values are not of interest in this research since the lower limits of the arcs are set equal to zero. It is impossible to devalue the lower bounds further.

An example may clarify this concept. Assume that in an optimal solution, power plant 1 (Figure 2.1) has a delivered coal price of $2 per ton, while power plant 2 has a delivered price of $4 per ton. The cost of transporting a ton of coal from power plant 1 to 2 (arc (1,2)) is $1 (c(1,2) = $1). By use of our definition of opportunity cost c_{12} = $1 + $2 - $4 = -$1. Therefore additional flow should be sent along arc (1,2). The -$1 represents the total savings for each additional ton of coal that is shipped to power plant 2 from plant 1 until either power plant 2 requires no more coal, or arc (1,2) can not carry any more coal.

If arc (1,2) has negative opportunity costs and is used to its full capacity in an optimal solution then, arc (1,2) is an arc if whose upper limits were increased would represent additional savings on transport costs. Arc (1,2) therefore represents an opportunity to further decrease cost, hence the term opportunity costs.

If on the other hand p(y) = $2 then $\overline{c_{12}}$ = +$1. In this case the flow along arc (1,2) should be reduced as much as possible. The value of +$1 means that each unit sent along arc (1,2) represents an +$1 per unit cost. Therefore to reduce total cost, the flow should be reduced as much as possible subject to the lower limit of the arc.

A value of zero for an opportunity cost indicates that there are no savings that can be gained by increasing flows along that arc. The use of opportunity costs can therefore identify those arcs in a distribution network that could significantly reduce total transport cost if upgraded to a higher upper limit.

Mathematically the problem solved by the out-of-kilter algorithm is:

$$\text{Min } Z = \sum_{(x,y)} c(x,y)\, f(x,y) \qquad (1)$$

subject to the constraints

$$\sum_{y} f(x,y) - \sum_{y} f(y,x) = 0 \qquad \forall x,y \qquad (2)$$

$$l(x,y) \leqslant f(x,y) \leqslant u(x,y) \qquad \forall x,y \qquad (3)$$

$$f(x,y) \geqslant 0 \qquad \forall x,y \qquad (4)$$

where Z - total cost of network flow
c(x,y) - cost per unit of flow of arc (x,y)
l(x,y) - minimum capacity of arc (x,y)
u(x,y) - maximum capacity of arc (x,y)
f(x,y) - flow along arc (x,y)

Equation 2 is a mathematical requirement to allow for solution, but it does not enter into the interpretation of the solution. It states that the inflow into a node must be equal to the outflow for that node. This requirement is easily met in the actual entry of the problem for computer solution.

In the end, the application of the out-of-kilter algorithm generates a solution that obeys the physical constraints of a distribution network while providing a minimization of transport costs. In addition, it can identify those arcs that present an opportunity to reduce transport cost through arc upgrading. It is then possible to compare the flows known to occur on a network in reality with those that should exist in an optimal allocation of flows. Discrepancies between the two indicate inefficiencies in the actual allocation of flows.

Comparison of arcs identified as having high opportunity costs with planned network modification allows for evaluation of the wisdom of such planned changes.

III
Changing Patterns in Soviet Natural Gas Flows

A. INTRODUCTION

Due to the various energy difficulties experienced by the USSR (and consequently CMEA) in the late 1970s, the Soviet leadership formulated a development program for the 11th Five-Year Plan (1981-1985) in which energy became the country's top industrial priority, with natural gas in the leading role. The plan ambitiously called for increasing the output of natural gas from 435 billion cubic meters (BCM) in 1980 to 630 BCM in 1985. Through 1984, increased gas output represented over 94 percent of the entire increment to overall energy supplies since 1980 (*Narkhoz SSSR 1984*, p.166).

The catalyst for this program was primarily the deteriorating situation for petroleum, the USSR's most important energy source. In 1980, petroleum supplied 45 percent of Soviet fossil fuel production and about 38 percent of consumption, while gas contributed about 27 percent and coal about 25 percent (Dienes, 1983a, pp.285-286; *Narkhoz SSSR 1980*, p.156). Despite its prominant role, Soviet petroleum prospects have become somewhat uncertain because of a deterioration in reserve position (Stern, 1981; Meyerhoff, 1983). Growth rates in crude petroleum production (which includes gas condensate) have slowed drastically since 1975 and production finally peaked in 1983; production in 1984 was 99 percent that of 1983, and production in 1985 was 97 percent that of 1984 (*Narkhoz SSSR 1984*, p.166; *Ekon. gazeta*, no.3, January 1986, p.2). Initially, the Soviets planned to offset this by shifting to coal but problems in the coal industry led to a decline in production in the late 1970s, forcing a change in strategy to gas (Gustafson, 1982).

Unlike petroleum, the development of gas is not hampered by lack of reserves, since the USSR possesses over 40 percent of the world's proven reserves of natural gas (Stern, 1983, p.363). Instead, the problem associated with natural gas is almost entirely one of geography. Over 75 percent of all gas reserves are located in Siberia, mostly in the northern part of Tyumen' Oblast in a few supergiant fields ("News Notes", SG-November 1983, pp.703-707). This places most of the USSR's gas potential roughly 3000 kilometers from the main industrial centers in the European USSR and about 4500 kilometers from the East European border for export (Figure 3.1). Therefore the feasibility of the gas plans would appear to rest largely upon the Soviet's ability to install the necessary pipeline carrying capacity to transport the gas to market.

Recent gas production trends highlight this problem. Whereas West Siberia supplied only 5.5 percent of Soviet gas in 1970, and only a modest 13.8 percent in 1975, by 1984 it supplied over 56 percent (Table 3.1). The Urengoy field, which yielded its first gas in 1978, produced over 200 BCM in 1984, or nearly 36 percent of Soviet output. This shift in production increased the average length of haul for gas from 1237 kilometers in 1975 to 2385 by 1983 (Brents et al., 1985, p.16).

Our purpose is to examine the ability of the Soviet Union to transport natural gas from distant new Siberian gas fields to the established industrial urban core in the European portion of the USSR as well as for export at the Western border. The future success of economic and industrial development in the USSR may well rest on the outcome. The ability and efficiency of the Soviet pipeline network to transport natural gas is studied by modeling the pipeline system as a capacitated network. Analysis is confined to the spatial dimensions of gas flows for 1970, 1975, 1980 and 1985.

All gas flows in the system are considered on an annual basis, although flows are evidently not constant during the course of the year (Sidorenko, et al., 1983, p.18). There are sizable seasonal fluctuations in demand, and the pipelines largely

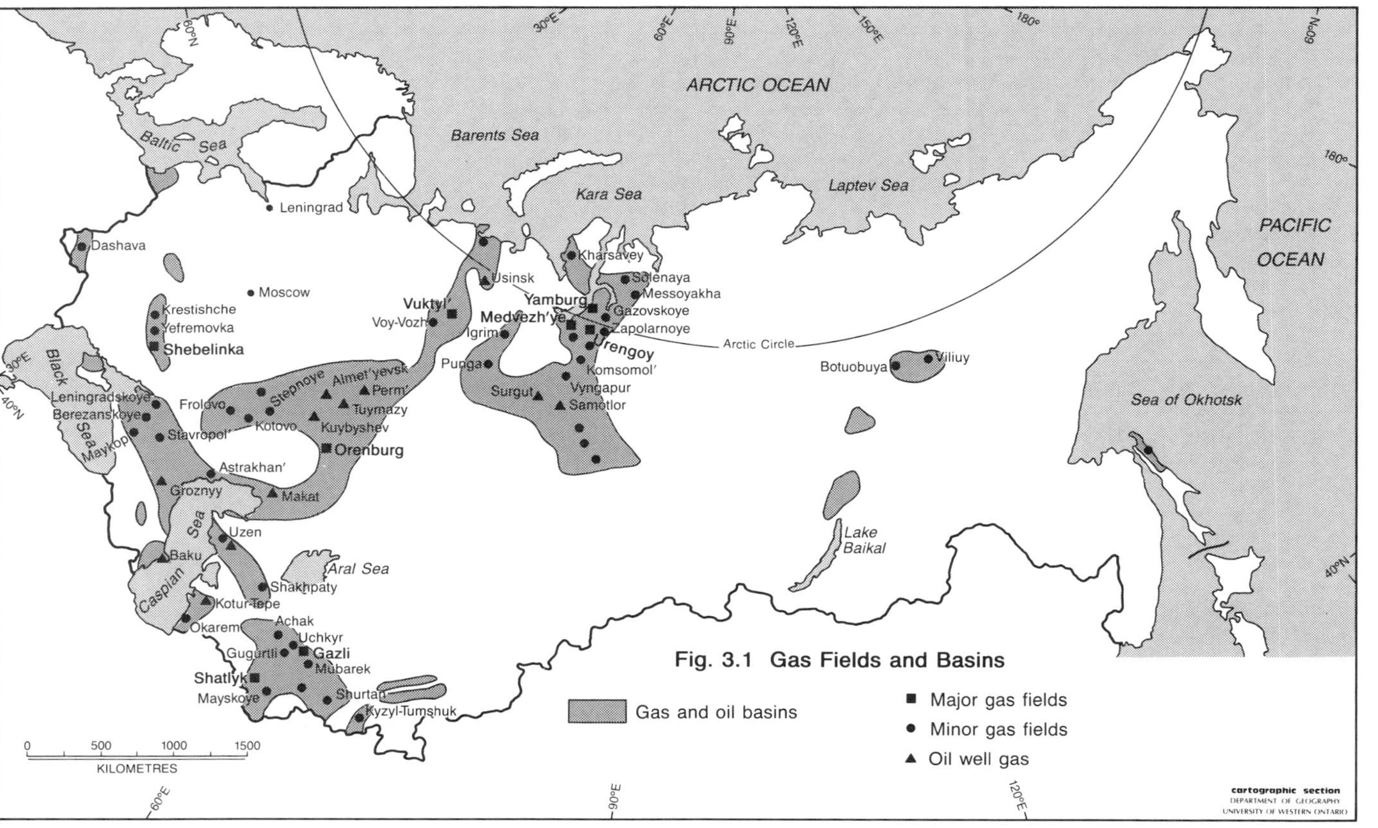

Fig. 3.1 Gas Fields and Basins

TABLE 3.1

Soviet Gas Production
(billion cubic meters)

	1970	1975	1980	1984
USSR	198	289	435	587
European Russia*	69	52.5	39.5	32
Komi ASSR	7	18.5	17.5	16
North Caucasus	47	23	14	9
Volga	15	11	8	7
Urals	3	23	52	50
Siberia	11	40	162	331
Tyumen'	9	36	156	324
Urengoy	--	--	50	210
Medvezh'ye	--	30	71	77
Vyngapur	--	--	16	18
Oilfield gas	0.1	1.7	12	19
Other Siberian	2	4	6	7
Ukraine	61	69	52	43
Azerbaijan	5.5	10	15	14
Kazakhstan	2	5	5	5
Uzbekistan	32	37	39	37
Turkmenia	13	52	70	75

Sources: "News Notes", SG-April, 1985, p.295; Dienes and Shabad, (1979, pp.70-71).

*Does not include the Bashkir ASSR.

are geared to peak winter demand because of insufficient underground storage capacity. The amount put into storage in 1980 was only 17.1 BCM, or only about 5 percent of annual consumption. In 1970 and 1975, the amount was even less, 5.4 and 14.2 BCM, respectively (Table 3.2). This places considerable stress upon the distribution system in winter. All gas demand cannot be met during the winter, so most gas-fired electrical stations are forced to use alternate fuels, particularly mazut (fuel oil) (Nekrasov and Troitskiy, 1981; Grigor'yev and Zorin, 1980), thereby undermining the policy of shifting consumption from petroleum to gas. Unfortunately, the two different seasonal flow regimes which evidently characterize the system cannot be modeled because of insufficient data.

B. THE GAS SYSTEM AS AN ABSTRACT NETWORK

In order to analyze the spatial dimensions of Soviet gas flows, the gas pipeline system first must be generalized into an abstract network of supply nodes, demand nodes, and bounded arcs. Supply nodes in the network are the major Soviet gas fields (Figure 3.1). Regional gas production data are available, in some cases to the field level ("News Notes", SG-April 1982, pp.283-287). For others, estimates from regional data must be used. A total of 43 supply nodes are included in the network in 1980, with fewer in the earlier years. The USSR also imports some gas, from Iran and Afghanistan (Stern, 1983, p.373). The supply nodes for these sources are located at Kelif, where the pipeline crosses from Afghanistan into Uzbekistan, and at Astara, where the pipeline from Iran enters Azerbaijan (Figure 3.2).

There are three types of demand nodes in the network. The first consists of foreign demand nodes for gas exports. Exports are allocated to four foreign demand nodes located at the Soviet border. The first, located north of L'vov, represents exports to Poland. The second, representing the largest amount, is located at Uzhgorod. It includes exports through the Bratsvo and Soyuz pipeline systems to Czechoslovakia, East Germany, Hungary, Yugoslavia, Austria, West Germany, Italy, and

TABLE 3.2

Budget for Pipeline Gas in the USSR
(BCM)

	1970	1975	1980
Receipts			
1. Gas production	199.6	291.0	435.2
2. Gas from storage	3.6	8.6	12.4
2. Gas imports	3.5	12.4	3.5
4. Total	206.7	312.0	451.1
Expenditures			
5. Gas to storage	5.4	14.2	17.1
6. Addition to amount in pipe	.1	.2	.5
7. Internal use of gas industry	7.2	16.4	31.2
8. Exports	3.3	19.3	55.6
9. Total	16.0	50.1	104.4
Consumption			
10. Domestic deliveries	190.7	261.9	346.7
11. Use by electrical stations	51.1	68.7	91.0
12. Deliveries to other industrial and municipal consumers	139.6	193.2	255.7

Sources:

Line 1: Ryps (1978, pp.12-13) for 1970, 1975; *Narkhoz SSSR 1980*, p.157 for 1980.

Lines 2,5: Ryps (1978, pp.12-13) for 1970, 1975; inflow for 1980 calculated by a residual from lines 6-9; outflow from Sidorenko *et al*., (1983, p.19).

Line 3: Ryps (1978, pp.12-13) for 1970, 1975; Stern (1983, p.373) for 1980.

Line 4: Sum of lines 1, 2, and 3.

Line 6: Ryps (1978, pp.12-13) for 1970, 1975; estimated for 1980 from length of gas pipeline (Stern, 1983, p.371) and 1975 ratio of amount to length, adjusted for the average increase in pipe size (see Sedykh and Kuchin, 1983, p.15).

Line 7: Ryps (1978, pp.12-13) for 1970, 1975, extrapolated for 1980 from the linear trend between 1970 and 1978.

Line 8: Ryps (1978, pp.12-13) for 1970, 1975; Stern (1983, p.373) for 1980.

Line 9: Sum of lines 5, 6, 7, and 9, except 1980 -- difference between line 4 and line 10.

Line 10: Line 4 minus line 9, except 1980. 1980 from Sidorenko *et al*., (1983, p.31).

Line 11: Ryps (1978, pp.12-13) for 1970, 1975; Nekrasov and Troitskiy (1981, p.224) for 1980.

Line 12: Line 10 minus line 11.

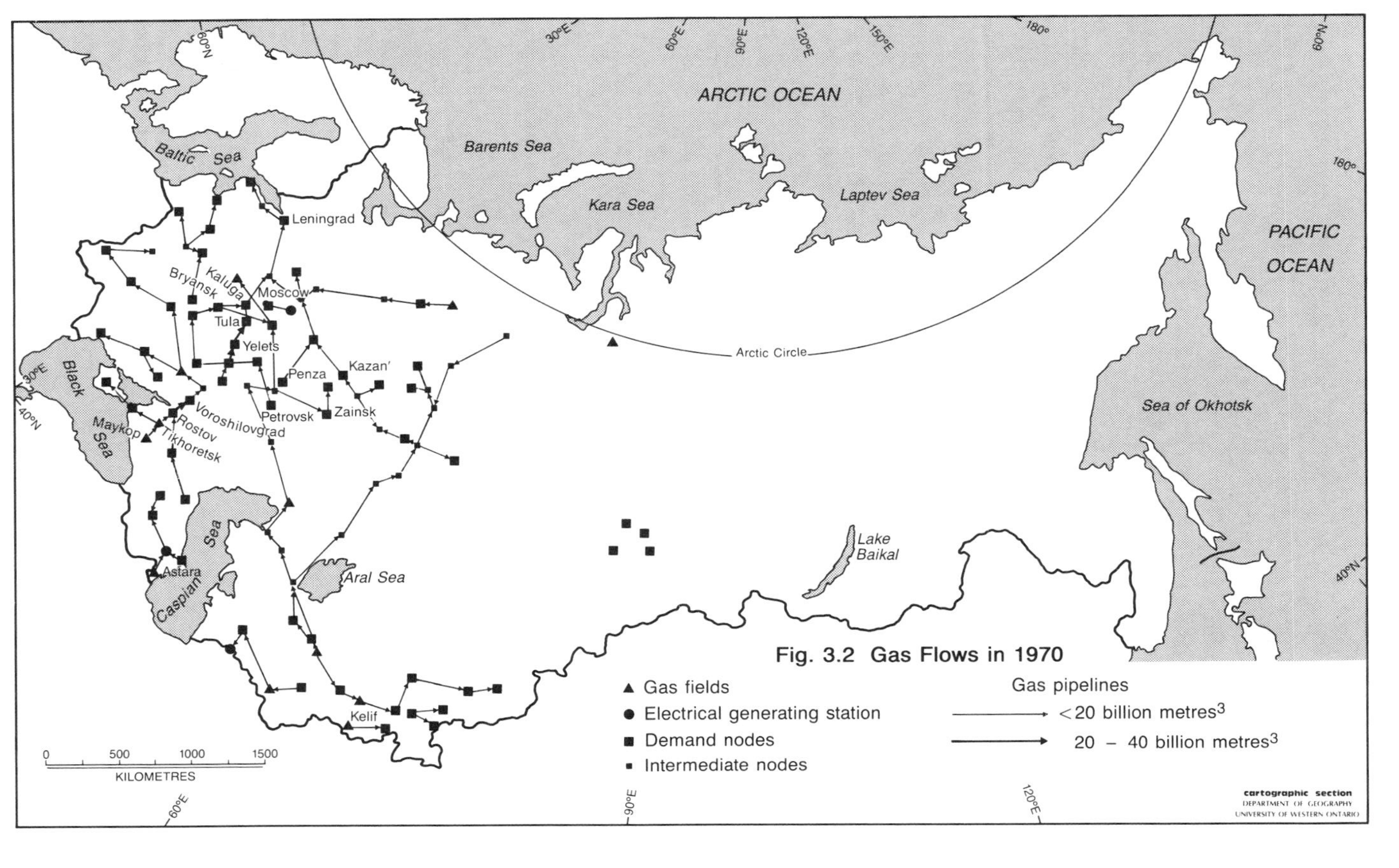

Fig. 3.2 Gas Flows in 1970

France. The third foreign demand node, for exports to Romania and Bulgaria, is located at Izmail. The fourth, at the Soviet-Finnish border, covers exports to Finland (Figure 3.2).

The other two types of demand nodes are for domestic consumption and consist of electrical stations and major cities. Here, domestic consumption is not considered to include the amount used internally by the gas industry itself (mainly to power compressor stations). This amount (31.2 BCM in 1980) is removed from the allocation (Table 3.2).

Electrical power stations are included as separate demand nodes because they are major consumers of natural gas in the USSR. In 1970 and 1975 they used 51.1 and 68.7 BCM, respectively, or about 26-27 percent of domestic gas consumption. In 1980, the Ministry of Electrical Power consumed 91 BCM in generating electricity, or 16 percent of domestic consumption (Table 3.2). Regional gas consumption data for electrical stations (Table 3.3) are allocated among the gas-fired electrical stations in each region according to installed capacity. Only major stations are considered in the analysis, which includes all those over 1000 MW and many smaller stations known to be at least 300 MW, for a total of 42 electrical stations.

Some regions contain only smaller TETs (heat and power stations) in the major cities, so in a few cases (e.g., Central Chernozem region), the amount utilized by electrical stations is allocated among the major cities in each region according to population size (see below), thus only implicitly accounting for consumption by TETs. For other regions which contain many smaller TETs and only a few major gas-fired stations (e.g., Urals, Volga, Ukraine), only half the regional consumption totals are allocated to the major power stations, with the other half allocated by city size to account for consumption at the smaller TETs.

Other domestic gas users are considered to be located in major cities. Principally these are industry (accounting for 56.7 percent of domestic consumption in 1975 or 148.4 BCM), and housing and the communal economy (accounting for 13.6 percent of

TABLE 3.3

Regional Gas Consumption by Electrical Stations (BCM)

	1970	1975	1980
USSR	51.1	68.7	91.0
Northwest	1.5	1.3	2.2**
Central	8.1	8.3	12.4
Central Chernozem	1.1	1.0	1.8**
Volga-Vyatka	.3	.6	1.3**
Volga	5.9	13.2	16.1*
North Caucasus	4.1	2.8	4.2
Urals	7.8	9.9	15.8*
West Siberia (a)	0	2.3	4.0
East Siberia (a)	0	0	0
Far East (a)	0	0	0
Baltic	.8	.8	1.3
Belorussia	.9	.9	.8
Ukraine	12.2	9.9	12.8
Moldavia	0	0	.4
Transcaucasus	1.1	6.0	6.7
Central Asia	5.2	9.4	8.9
Kazakhstan (a)	2.1	2.3	2.3

*Only half of this amount is actually allocated to the power stations in the region to adjust for consumption by TETs (heat and power stations).

**There are no major stations in the region (only TETs) so the total is allocated among the demand nodes in the region according to city size.

Sources: Data for 1970 and 1975 for most regions are from Ryps (1978, p.38). The amounts for regions marked (a) have been disaggregated from Ryps' residual "other" regions category according to relative capacities of major gas-fired stations in the various regions. Total consumption in 1980 is from Nekrasov and Troitskiy (1981, p.224). Estimates for the regions in 1980 are from data in Nekrasov and Troitskiy (1981, p.232), Fedorov *et al.*, (1981, p.7), and Dienes and Shabad (1979, pp.70, 93). The percentages given in Fedorov *et al.* leave no room for consumption in Kazakhstan; since the 1230 MW of gas-fired capacity at Dzhambul still existed, it was assigned the same amount as in 1975. Some regional estimates are disaggregated from larger regional groupings according to relative capacities of gas-fired stations.

consumption) (Ryps, 1978, p.13). Within industry, the largest consumer is ferrous metallurgy, accounting for nearly one-quarter of industrial consumption in 1975 (34.4 BCM), with other significant amounts being used in the chemicals, machine-building, and construction materials industries (Ryps, 1978, p.13).

Data on the regional distribution of gas consumption for these users are given in Table 3.4. The amount in each region is allocated among the major cities in each according to population size (Narkhoz SSSR 1979, pp.18-28). This approximately reflects the relative size of their industrial bases and communal economies (Huzinec, 1978). "Major" cities are considered to include all those on the network over 300,000 by 1980, as well as all oblast centers and some smaller cities located at key points in the gas network (e.g., Kotlas, Yelets, Serov, and Serpukhov). A total of 144 cities are included as demand nodes in 1980. Fewer are included in the earlier years because of the smaller network. Moscow is the largest consumer, demanding 25.0 BCM of gas in 1980 plus another 4.7 BCM for its 3500 MW TETs. This represents nearly 9 percent of domestic consumption. Leningrad's demand in 1980 is about half this at 12.2 BCM. In contrast, a number of cities (e.g., Kemerovo, Novosibirsk, Gulistan, Krasnovodsk, Kotlas) have consumption estimated at only 0.1 BCM in 1980.

The only exception to the method of allocating the regional total among the cities is the Ukraine. Consumption data for Ukrainian cities in 1970 and 1975 are given in Kalchenko et al. (1974, p.42). However, these data must be adjusted by subtracting the amount allocated to the major electrical stations (estimated above). Not surprisingly, these data show much higher consumption of gas in the large cities of the eastern Ukraine where ferrous metallurgy predominates (e.g., Donetsk, Voroshilovgrad), when compared with an allocation based upon just population size. This probably means that consumption for other centers of ferrous metallurgy, such as Magnitogorsk and Cherepovets, is underestimated.

TABLE 3.4

Gas Deliveries to Industrial, Municipal and Other Users
(Excluding Electrical Stations)
(BCM)

	1970	1975	1980
USSR	139.6	193.2	255.7
Northwest	7.8	11.4	18.4
Central	22.0	26.8	38.8
Central Chernozem	3.5	5.1	6.1
Volga-Vyatka	4.4	4.5	5.6
Volga	15.6	18.1	24.3
North Caucasus	10.7	14.2	16.4
Urals	17.2	25.7	30.8
West Siberia	.4	.8	6.0
East Siberia	.4	2.6	3.8
Far East	1.2	1.3	1.4
Baltic	2.0	3.0	4.5
Belorussia	2.0	2.6	3.7
Ukraine	36.7	47.2	59.2
Donets-Dnieper	22.5	28.6	35.9
Southwest	12.0	16.3	20.6
South	2.2	2.3	2.7
Moldavia	.2	.4	.6
Transcaucasus	7.4	10.9	13.9
Central Asia	6.6	12.1	15.1
Kazakhstan	1.5	6.2	7.0

Sources: Data for 1970 and 1975 are from Ryps (1978, pp.15, 38). The Ukrainian data are from Kal'chenko *et al*. (1974, p.42), although adjusted to sum to Ryps' figures. 1980 data are estimated from Fedorov *et al*. (1981, p.7), Sidorenko *et al*. (1983, p.31) and the estimates in Tables 3.2 and 3.3.

In addition to these supply and demand nodes, 41 intermediate nodes are added to the network to allow for junctions or changes in capacities at sites other than supply or demand nodes. These include such key junction points as Torzhok, Ostrogozhsk, Novopskov, Aleksandrov-Gay, Nadym, and Gryazovets.

Bounded arcs represent the pipelines. The capacity of each arc is set by an upper bound determined by the reported operating average by size of pipe (Dienes and Shabad, 1979, p.83; CIA, 1978, Appendix I), ranging from 1.5 BCM annually for a 530 mm pipe to 25.0 BCM for a 1420 mm pipe. The newly constructed 1420 mm pipelines reportedly carry 30-32 BCM (Hewett, 1984, p.77). However, several other factors, most importantly compressor capacity, affect actual carrying capacity. Thus not all pipelines, or evidently, even particular sections of the same line, operate at these levels. Delays in installing compressor capacity are known to plague the Soviet gas industry, resulting in many pipelines operating under design capacity (Stern, 1983, p.369). However, these factors could not be incorporated in the model because of insufficient data.

Distance is used as a surrogate for transportation cost in the network since transportation costs in the Soviet gas system are largely determined by the distance gas has to be shipped, although other factors such as the size of pipe and the number of strings also affect this (Tretyakova, 1985). Distances in this and subsequent models are calculated from the latitude and longitude of the terminal nodes for each arc. The length of an arc thus becomes the cost of utilizing it.

C. RESULTS: 1970, 1975, AND 1980

The modeled flows for 1970 are shown in Figure 3.2. The model-generated flows correspond quite closely to known actual flows (Gankin *et al*., 1972, p.100). The small size of the flows indicates a relatively underutilized energy source as well as a somewhat skeletal distribution system. This corresponds with the noted immaturity of the industry as well as its relatively late start (Dienes and

Shabad, 1979). For example, the main flow in the system, from the North Caucasus to Central Russia, is only 21.3 BCM (this is between Rostov and Voroshilovgrad), and exports to Eastern Europe are only 3.3 BCM.

The model provides an opportunity cost for each of the arcs as part of an optimal solution. Negative opportunity costs allow for identification of arcs that are fully utilized and thus serve as constraints on the distribution of gas. The solution for 1970 has only a few significant constraining arcs. The scattering of the locations of the arcs indicates that no one region is underbuilt and responsible for inefficiency in the gas distribution system. The five arcs with the largest opportunity costs (and therefore present the options for the greatest savings in transport costs) are Petrovsk-Penza, Zainsk-Kazan', Maykop-Tikhoretsk, Bryansk-Kaluga, and Yelets-Tula (Figure 3.2). These constraints arise mostly along the main route from the North Caucasus to Moscow, although two are connecting lines in the Volga region.

The 1975 pattern of flows is shown in Figure 3.3. The flow pattern becomes somewhat more complex, with a greater number of utilized pipelines as well as larger flows. A significant new field at Medvezh'ye in Tyumen' Oblast is utilized. Substantially larger flows now take place from Central Asia to Central Russia through Beineu, Makat, and Aleksandrov-Gay (Figure 3.3). This reflects the role of Central Asia as an intermediate producing region after the early limited resources of the European USSR in the North Caucasus and Ukraine had been developed but before the vast potential of West Siberia could be fully exploited (Dienes and Shabad, 1979, p.79). Thus the region reached its apogee in terms of importance in the mid-1970s.

The flow map corresponds quite closely with known flow and supply patterns for 1975 (Furman, 1978, p.39). The hypothetical flows also reflect the relative change in total cost (BCM-kilometers) for the Soviet system from 1970 to 1975. Costs rose in the modeled network from 146.9 BCM-kilometers to 292.6 BCM-kilometers from 1970 to 1975, an increase of 50.2 percent. In the same period actual costs

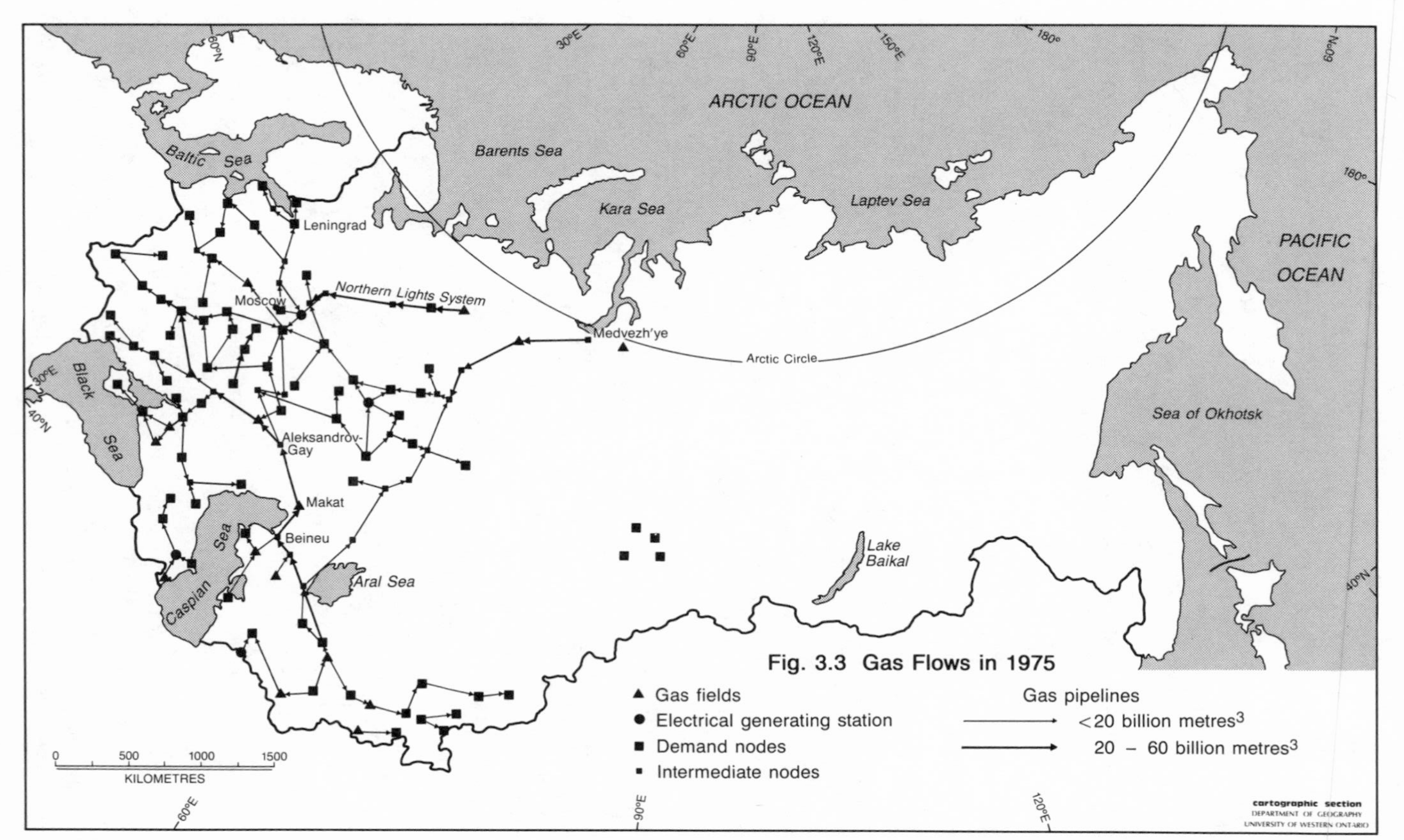

Fig. 3.3 Gas Flows in 1975

rose by 48.2 percent, from 166.4 to 345.6 BCM-kilometers (Stern, 1983, p.371).

The increasing strain being placed on the gas distribution system noted by Soviet sources is indicated by the number of arcs being utilized at or above stated capacities. In 1970, 16 arcs in the model are at upper capacity, while in 1975 the number more than doubles to 37. One set of arcs comprising the Northern Lights system (Figure 3.3), is utilized at a level considerable above the 20 BCM of capacity originally assigned. The high opportunity costs associated with these arcs indicate an upgrading of the pipelines would increase efficiency. This result correctly anticipates the actual upgrading of the Northern Lights system's capacity to 56 BCM by 1980 (CIA, 1978, p.59).

The major modeled flows for 1980 (Figure 3.4) generally correspond with Soviet projections of flow patterns for 1980 (Furman, 1978, p.40). The hypothetical flows also replicate the relative change in total BCM-kilometers in the Soviet system for this period. For example, costs rose in the modeled network from 292.6 BCM-kilometers in 1975 to 624.2 in 1980 (47 percent). In actuality, costs rose by 47 percent, but from 345.6 to 742.3 BCM-kilometers (Stern, 1983, p.371). Modeled costs are only 85 percent of actual BCM-kilometers due to the abstractions used, such as straight line distances and imperfections in the demand estimates. However, the results from the model do imply that the Soviet system is operating near optimum efficiency.

One interesting result is that the model generates almost no flow through the Central Asia-Urals pipeline in 1980, as the line becomes superfluous with the advent of Siberian gas after 1975. This does occur in reality, as the empty pipeline is being considered as a conduit for transferring water from West Siberia to Central Asia (Kibalchich, 1983).

The 1980 allocation shows an increase in the amount of congestion in the system. Over 60 arcs are at their upper bounds, and several major problem areas are highlighted from adjustments necessary to achieve a feasible solution. One, the main line

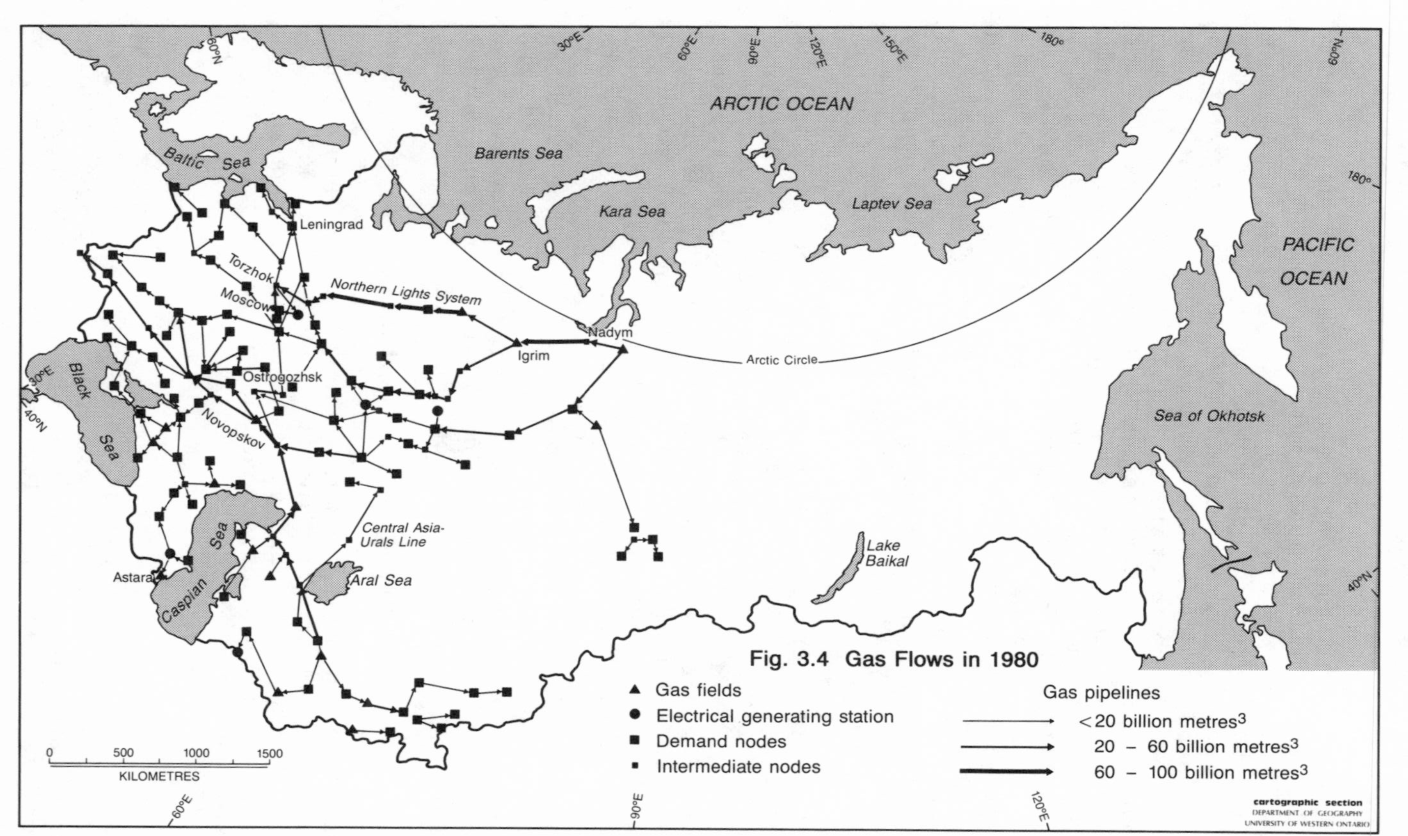

Fig. 3.4 Gas Flows in 1980

from West Siberia, has a severe lack of transmission capacity. This arc, between Nadym and Igrim (Figure 3.4), connects the giant West Siberian fields with the Urals and Northern Lights transmission systems and is required to transmit 90 BCM, about 30 percent more than its original assigned capacity. As in the 1975 solution, the other arcs in the Northern Lights system that transverse the Komi ASSR to Torzhok (Figure 3.4) are also either exhibiting flows at or in excess of originally assigned upper capacities. This implies that this line requires further upgrading for the efficient distribution of natural gas and confirms the view that additional pipeline construction is necessary for the success of the gas plans.

Another major bottleneck is in the main lines feeding gas from Central Asia, West Siberia, and Orenburg into the major distribution points at Ostrogozhsk and Novopskov (Figure 3.4). They are forced to operate well over assigned upper bounds in order to meet the large demand for eastern gas in the region caused by the decline of production in the Ukraine and North Caucasus.

The third major area experiencing difficulties is the Transcaucasus. The problem here is due to supply disruptions following the revolution in Iran. To offset the anticipated decline in local production, the USSR began to import gas from Iran in 1970 to supply the three Caucasian republics. Imports reached 9.6 BCM in 1975, but following the Iranian revolution, imports dropped to about 1.0 BCM in 1980 before ceasing altogether (Stern, 1983, p.373). This causes a local gas shortage since the lack of transmission capacity precluded bringing in alternate supplies. The magnitude of this problem is illustrated by the fact that the gas source at Astara has to be increased from 1 to 5 BCM to make the solution feasible. This can be taken as a measure of the supply shortfall in the region.

These two last bottlenecks illustrate some of the problems being experienced by the older gas-producing areas as their supplies are depleted, necessitating the reorientation of gas flows through existing lines.

D. 1985 SCENARIO

The success at modeling Soviet gas distribution for 1970, 1975, and 1980 allows us to present an analysis of probable flow patterns for 1985. This is accomplished by adding several new lines to the system and increasing supply and demand at several key nodes.

The 11th Five-Year Plan (FYP) called for the entire increase in gas supplies to come from the supergiant field at Urengoy in West Siberia. To accommodate this increase, six large pipelines from the field were slated for construction between 1981 and 1985, following existing routes. So far, four of them have been more or less completed, with line work on the fifth and sixth being completed in late 1984 and early 1985 ("News Notes", SG-April 1983, p.318; December 1983, pp.775-777; "News Notes", SG-April 1985, p.296). Although the installation of compressor capacity is lagging, it is likely that at least the first five lines will be fully operational in 1985.

One pipeline follows the Northern Lights system through Ukhta to Moscow (and continues on through Torzhok and Minsk to the export point at Uzhgorod), thus alleviating some of the congestion on the Northern Lights system noted in the 1980 allocation (see above). It is interesting to note that it was the first pipeline constructed in the 11th FYP Plan, becoming operational in 1981.

Two of the other new lines are intended to deliver gas to key junctions for further distribution. One pipeline runs from Urengoy through Perm' to Petrovsk, while another follows this same route but continues on from Petrovsk to Novopskov. The line to Petrovsk became operational in 1982, and the line to Novopskov, in 1983. These lines should alleviate the capacity limitations indicated at Ostrogozhsk and Novopskov in the 1980 solution (see above). An extension of this second line was constructed in 1983 and 1984 from Novopskov to Kazi-Magomed in Azerbaijan to offset the loss of imports from Iran ("News Notes", SG-April 1984, p.275).

A fourth pipeline runs from Urengoy to Uzhgorod for export. Pipelaying was completed on this controversial export line in 1983, and all compressor capacity was due to be installed by mid-1984 ("News Notes", SG-December 1983, pp.775-777). The fifth pipeline, running from Urengoy through Perm' to the central area of the European USSR (Yelets/Khar'kov), was completed in 1984. Line work on the sixth pipeline, also running from Urengoy to the central European USSR (Yelets/Kiev), was completed in February 1985.

Five of the new pipelines are added to the existing (1980) network and assigned a capacity of 25 BCM, with demand increased at the terminii of each a like amount. Thus demand at Moscow, Petrovsk, and Khar'kov is increased by 25 BCM, while demand at Novopskov and Kazi-Magomed (the two terminii of the third line) is made 15 and 10 BCM, respectively. Exports through Uzhgorod are set at 70 BCM, 30 BCM of which represents exports to Western Europe. Therefore in the scenario, total Soviet exports are 82 BCM. Exports were already nearly 70 BCM in 1984 (Osipov, 1985, p.8).

The addition of the five new lines in the model allows gas production at Urengoy to be increased by 140 BCM, so the scenario places "useful" Soviet gas production in 1985 at 575 BCM, with 190 BCM being produced at Urengoy. To this can be added the amount consumed by the gas industry itself in 1980 (Table 3.2), as this was removed from the analysis. This would make total production in 1985 over 600 BCM, of which 221 BCM would be at Urengoy. Actual production in 1983 was 536 BCM, and that of 1984 was 587 BCM (Ekon. gazeta., January 1985, p.6). Production in 1985 was 643 BCM (Ekon. gazeta., no.3, January 1986, p.2), mainly because of the early completion of the sixth pipeline.

The projected pattern of flows for 1985 (Figure 3.5) is very similar to that of 1980, although the flows from Western Siberia are much larger. This similarity indicates that the network and the pattern of flow are reaching long run stability. Previously the demand for natural gas in the Soviet Union was relatively fixed in the European portion of the country due to the inertia of the existing

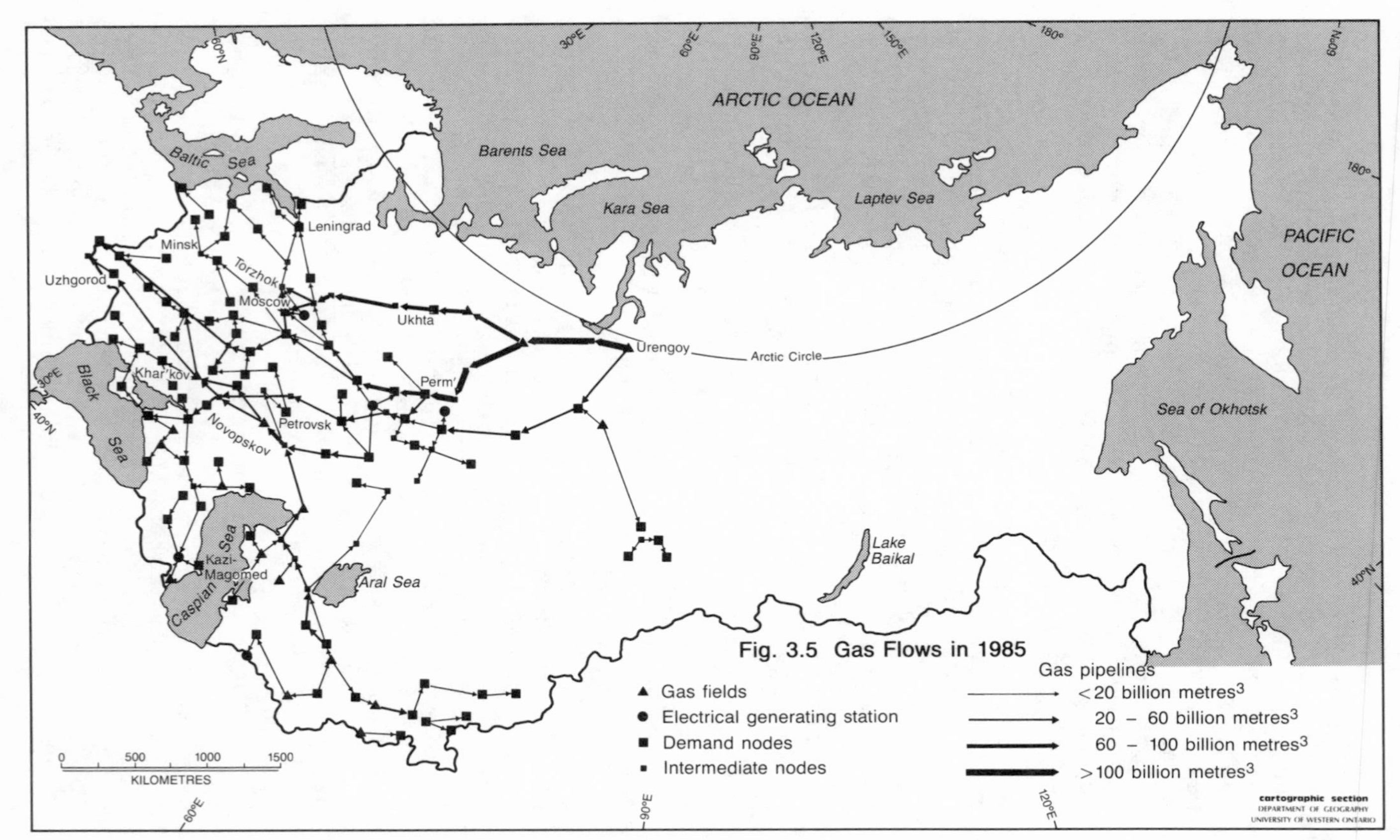

Fig. 3.5 Gas Flows in 1985

industrial-urban spatial pattern while the supply of natural gas was shifting farther into the hinterland. This pattern occurred because the reserves of natural gas of the more accessible fields in the Ukraine, North Caucasus, and Central Asia were comparatively modest relative to long-run demand. Thus a geographically stable relationship between supply and demand has been realized with the opening of the giant Siberian fields. The model indicates that this increased flow of gas from West Siberia will increase total system cost approximately 65 percent over that in 1980.

The introduction of these new pipelines increases the overall efficiency of the distribution network. The number of arcs being utilized at full capacity decreases to 48 and the opportunity costs for all arcs become relatively low. Spot upgrading of lines will no longer significantly reduce costs. The rather significant increase in exports contemplated between 1980 and 1985, including those to Western Europe, does not particularly tax the gas system with the additional lines installed.

E. CONCLUSIONS

The bottlenecks identified in the analysis seem to correctly anticipate subsequent developments. Flow patterns generated by the model replicate what little is known about actual flows. Changes in relative costs are also closely duplicated, although actual totals vary because of model assumptions. This implies that the Soviet gas system is operating near optimum efficiency. Once the Soviet gas system is modeled in its basic dimensions, a scenario for 1985 is examined. The effect of an additional five large pipelines from the West Siberian field at Urengoy is determined. The addition of such pipelines increases efficiency and alleviates the most serious constraints in the system.

In more general terms, this analysis has moved research from aspatial estimates of supply and demand for natural gas to the very real problem of distribution. For the first time the gas distribution system has been examined in a holistic, although somewhat abstract form. This study represents an advance over previous reports of specific bottlenecks examined in isolation. An interesting extension of the analysis would be to determine if the pace of construction of compressor stations is related to bottlenecks in the flow of gas.

Now that the basic structure of the Soviet natural gas pipeline network has been successfully modeled, additional scenarios could be examined. Inclusion of seasonal flow fluctuations and expansion of underground storage facilities could be undertaken as well as the integration into the model of construction delays and new or proposed pipelines. Such a study would allow for a sensitivity analysis of alternate routes and would provide a measure of the rationality of suggested changes.

IV
Transport Constraints in Soviet Petroleum

A. INTRODUCTION

Although the USSR has been the world's leading producer of crude petroleum since the mid-1970s, the Soviet oil situation has become somewhat uncertain in recent years because of a deterioration in reserve position (Stern, 1981; Meyerhoff, 1983; Gustafson, 1985). As a result, growth rates in crude production (which includes gas condensate) have slowed drastically since 1975 and production finally peaked in 1983 at 616.3 million tons; production in 1984 was slightly less, 612.7 million tons, and production in 1985 was only 595 million (Narkhoz SSSR 1984, p.166; Ekon. gazeta., no.3, January, 1986, p.2).

While the most serious long term problem in Soviet oil prospects probably concerns the deteriorating reserve position, the situation is exacerbated by a transportation problem because of the changing geography of petroleum production. Before World War II, Soviet oil production was concentrated in the Caucasus, at Baku and Groznyy. In the mid-1950s the center of production shifted to the Volga-Urals region, and output grew rapidly so that by the early 1960s the USSR was the world's second largest producer after the United States. Then in the 1970s, the Volga-Urals fields were surpassed in production by those of West Siberia. By 1980 over half of Soviet crude was coming from West Siberia and now the proportion is over 60 percent (Figure 4.1) ("News Notes", SG-April 1985, p.289). In contrast, West Siberia accounted for only about 9 percent of Soviet oil output in 1970 (Table 4.1).

In this chapter, the efficiency of transporting crude petroleum in the USSR is examined. Although difficult because of limited information on oil

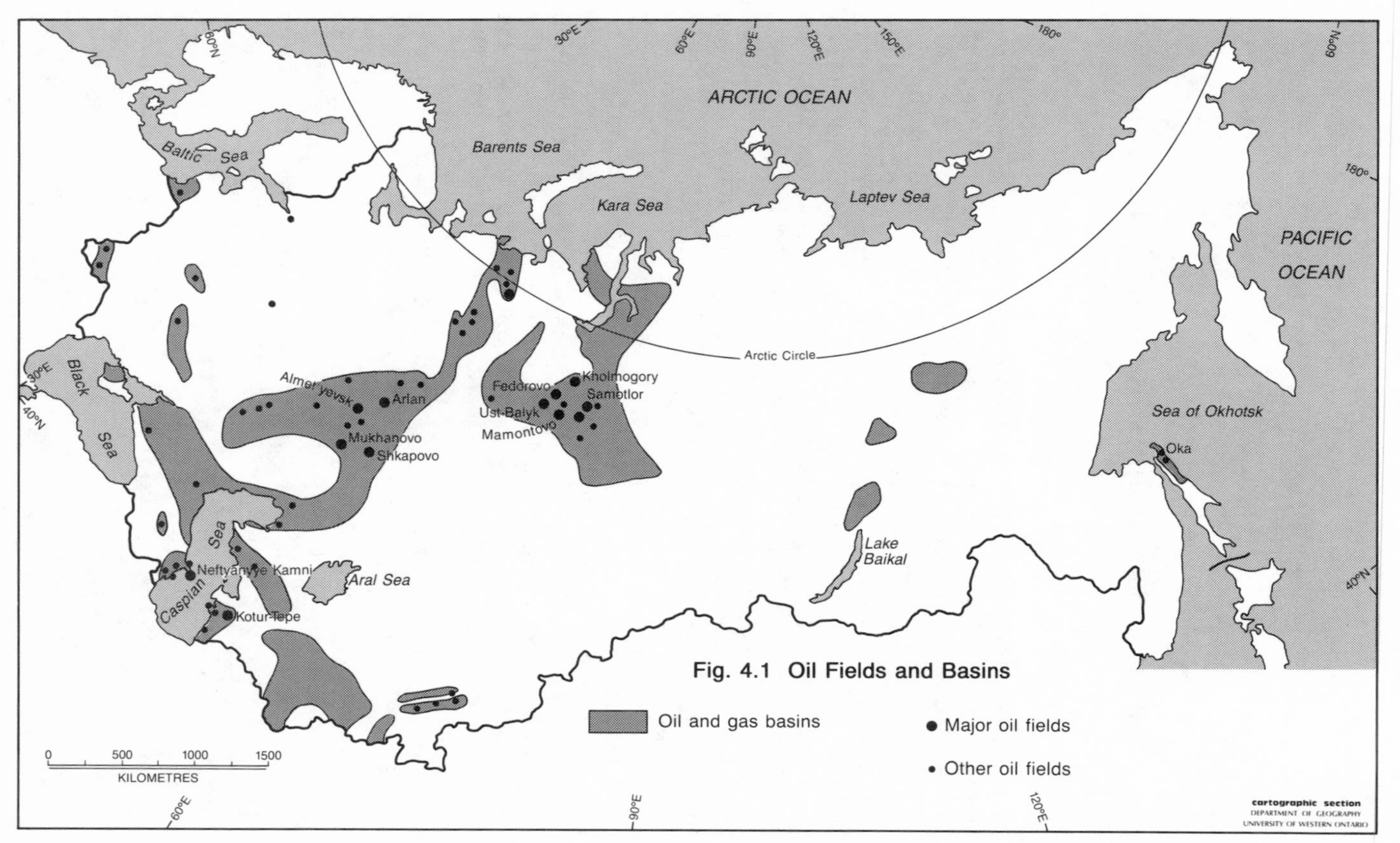

Fig. 4.1 Oil Fields and Basins

TABLE 4.1

Soviet Oil Production*
(million tons)

	1970	1975	1980	1984
USSR	353	491	603	613
European Russia	188	182	149	107
Komi ASSR	8	11	19	19
North Caucasus	35	24	19	12
Volga	145	147	111	76
Urals**	63	79	82	73
Siberia	34	151	316	381
West Siberia	31	148	313	378
Sakhalin	2.5	2.5	3	3
Ukraine	13.9	12.8	7.6	n.d.
Belorussia	4.2	8	3	n.d.
Georgia	0.02	0.3	3.2	n.d.
Kazakhstan	13.1	23.9	18.7	20.6
Central Asia	16.8	17.7	9.5	n.d.
Azerbaijan	20.2	17.2	14	n.d.

Sources: "News Notes", SG-April, 1985, p.289, April 1985, p.289, Dienes and Shabad (1979, pp.46-47).

*Includes gas condensate.

**Includes the Bashkir ASSR.

production, refinery capacities, and oil pipelines, Soviet petroleum movements are analyzed with fragmentary data, identifying constraints and future problem areas.

The Soviet petroleum distribution system as it existed in 1980 (Figure 4.2) is also modeled as a capacitated network to generate optimal hypothetical flows for the system. The purpose is not to replicate reality in detail, but to indicate general patterns, as relatively little is known about the overall magnitude or direction of actual petroleum flows in the Soviet Union.

B. SIBERIAN PIPELINE DEVELOPMENT

The development of the new West Siberian fields presented considerable difficulties, as the swampy, forested region along the middle reaches of the Ob' River, where most of the fields were located, was largely unsettled with little existing infrastructure. Furthermore, the West Siberian fields were 1500 miles further east than those of the Volga-Urals, and that much further from existing refineries and the main consuming centers in the European USSR.

To move oil from the rapidly expanding new fields in West Siberia to the main refining centers in the European USSR required the construction of a massive new pipeline network. But as the development of the West Siberian fields began in the mid-1960s, crude was initially moved by tanker barge from the fields near Surgut and Nizhnevartovsk down the Ob' and then up the Irtysh River to Omsk for refining. This was a stopgap measure, and as the volume of production grew, a 1020 mm pipeline was constructed in 1967 to Omsk, joining the existing trans-Siberian pipeline network. By 1970, West Siberian production had grown to the point where it exceeded the capacity of the Siberian refineries at Omsk and Angarsk, so the flow in the existing trans-Siberian pipelines from Ufa was reversed, supplying Siberian crude to the Volga refineries for processing (Dienes and Shabad, 1979, p.57).

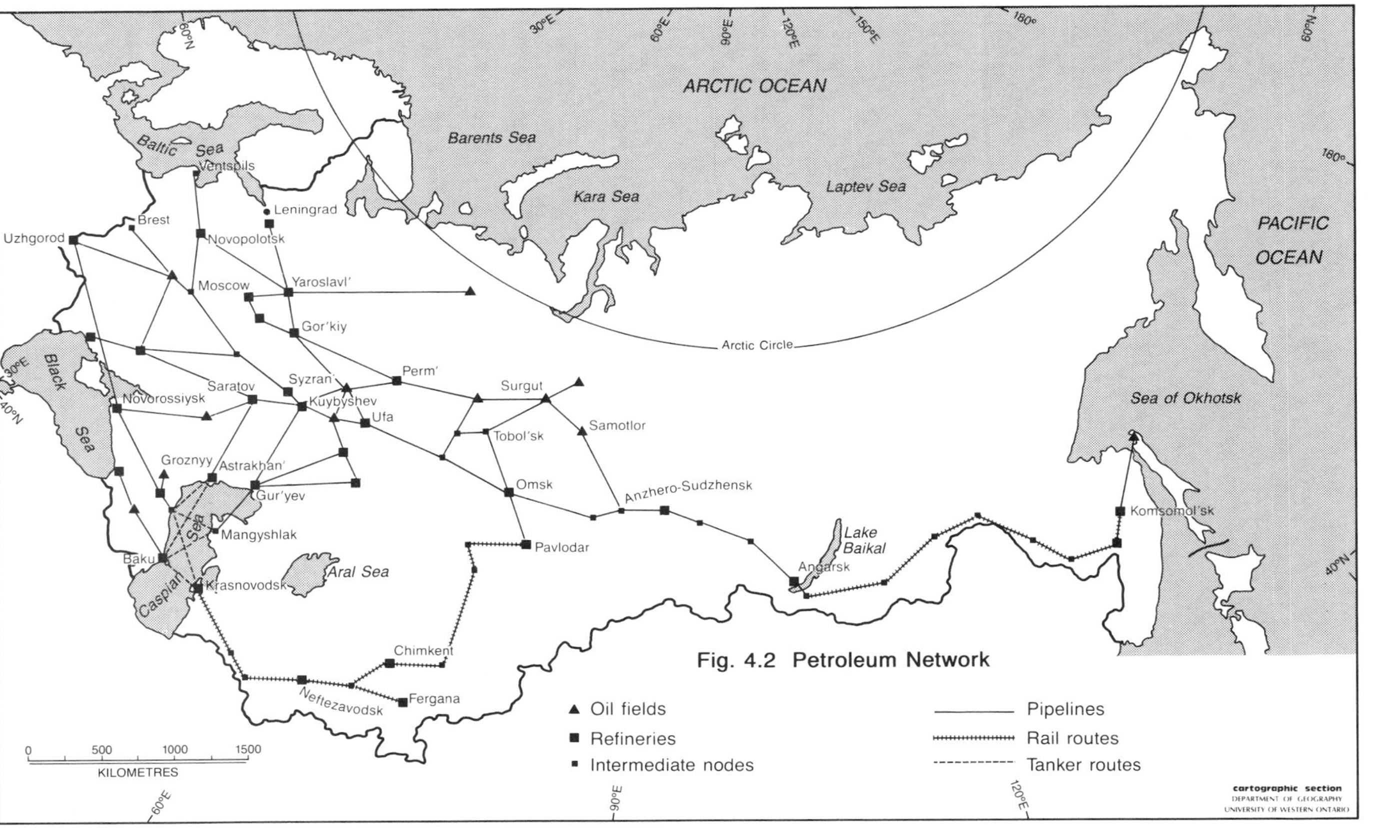

Fig. 4.2 Petroleum Network

The rapid growth of West Siberian production, associated with the development of the giant Samotlor field in the early 1970s, required further expansion of the pipeline network. In 1972 a 1220 mm line from Samotlor east to Anzhero-Sudzhensk was completed (Figure 4.2), and a year later it was extended to Krasnoyarsk (Dienes and Shabad, 1979, p.67). Apparently by 1981 the line had reached the Angarsk refinery near Irkutsk (Shcherbina _et al._, 1981).

Two other 1220 mm lines were completed in the 1970s to the western USSR. The first, the Samotlor-Almet'yevsk line, was finished in 1973, and the second, between Nizhnevartovsk and Kuybyshev, in 1976 (Dienes and Shabad, 1979). Both follow a southerly route through Tobol'sk, Tyumen', and Kurgan (Figure 4.2). These new lines increased transmission capacity between West Siberian and the European USSR to about 175 million tons per year (for comparison, total Soviet production in 1975 was 491 million tons).

In 1977 an 820 mm line was constructed south from Omsk to the new refinery at Pavlodar in Kazakhstan, and by 1983 it had been extended to Chimkent, the site of a refinery then under construction. Plans call for the eventual extension of this pipeline to another new refinery under construction at Neftezavodsk, near Chardzhou in Turkmenia, with a spur to supply the existing refineries at Fergana and Khamza in Uzbekistan.

In the late 1970s, construction began on a 1220 mm pipeline from Surgut to Novopolotsk in Belorussia, following a more northerly route to supply refineries at Perm', Gor'kiy, Yaroslavl', and Novopolotsk (Figure 4.2). The line had been extended as far as Yaroslavl' by 1980, and was completed in 1981 ("News Notes", _SG_-April 1982, p.278), adding another 75 million tons to westward transmission capacity. However, work on the Surgut-Novopolotsk pipeline was not completely finished until August 1984 (Foreign Broadcast Information Service, 8 August 1984, p.4).

Other adjustments in the pipeline network have been required as petroleum production has increasingly shifted to West Siberia. Additional trunk lines were constructed in the Ukraine, North Caucasus and Azerbaijan to handle the new pattern of flows ("News Notes", SG-October 1983, pp.625-627; April 1978, p.275; February 1975, p.122).

C. THE PETROLEUM DISTRIBUTION SYSTEM AS AN ABSTRACT NETWORK

To analyze the spatial dimension of petroleum flows, the out-of-kilter algorithm (OKA) is applied to the oil distribution system. This network consists of supply nodes (oil fields), demand nodes (oil refineries and export points), and bounded arcs (pipelines and other transport links).

Identifying supply nodes and supply availability is straightforward, as they comprise the major Soviet oil fields and/or oil-producing regions. A total of 31 supply nodes are utilized, producing 603 million tons of petroleum (Figure 4.1 and Table 4.2).

There are two types of demand nodes in the network, domestic and foreign. Virtually all crude is refined before going to final consumers as various refined products such as mazut, diesel fuel or gasoline, so the domestic demand nodes are the petroleum refineries (or refinery complexes if more than one plant is located in a city). In 1980, there were 42 of these refinery locations. Primary refinery throughput for each has been estimated previously, providing domestic demand estimates for crude petroleum (Table 4.3). Since gas condensate production is included in Soviet petroleum production figures, but is not included in the estimates of refinery throughput (Sagers, 1984, p.10), a dummy demand node for gas condensate (18 million tons in 1980) is included in the network near its center at Kuybyshev.

Foreign demand nodes represent petroleum exports, which totalled about 119 million tons in 1980 (of which 73 million went to the Eastern

TABLE 4.2

Petroleum Supply Nodes

Field/Region	Location	1980 production* (million tons)
USSR		603
Surgut	Surgut	46
Samotlor	Nizhnevartovsk	203
Ust-Balyk	Nefteyugansk	46
Sosino-Sovetskoye	Strezhevoy	5
Kholmogory	Noyabr'sk	3
Uray (Shaim)	Uray	6
Vasyugan	Srednyy Vasyugan	5
Perm'	Chernushka	23
Bashkir	Tuymazy	17
Bashkir	Neftekamsk	17
Tatar	Almet'yevsk	82
Orenburg	Buguruslan	13
Kuybyshev	Otradnyy	13
Kuybyshev	Neftegorsk	13
Mangyshlak	Uzen'	14
Emba	Makat	4
Turkmenia	Kotur-Tepe	8
Azerbaijan	Baku	14
Chechen-Ingush	Groznyy	7
Eastern Ukraine	Chernigov	6
Western Ukraine	Ivano-Frankovsk	2
Belorussia	Rechitsa	2
Komi	Usinsk	20
Stavropol'	Neftekumsk	6
Krasnodar	Neftegorsk	4
Dagestan	Yuzhno-Sukhokumsk	2
Udmurt	Arkhangel'skoye	8
Other Volga	Volgograd	4
Georgia	Tbilisi	3
Fergana	Andizhan	1
Sakhalin	Okha	3

Sources: "News Notes," April 1982, pp.177-283; April 1984, p.267; Dienes and Shabad (1979, Chapter 3).

*May not add to total because of rounding in estimates.

TABLE 4.3

Domestic Petroleum Demand Nodes

Refinery/City	1980 Throughput (million tons)	Refinery/City	1980 Throughput (million tons)
Achinsk	0	L'vov	1
Angarsk	23	Mazeikiai	7
Baku (three plants)	25	Moscow	10
Batumi	5	Mozyr'	12
Chimkent	0	Nadvornaya	2
Drogobych (two plants)	2	Neftezavodsk	0
Fergana	7	Nizhnekamsk	8
Gor'kiy	25	Novopolotsk	23
Groznyy	15	Odessa	3
Gur'yev	6	Omsk	26
Ishimbay-Salavat (two)	10	Orsk	4
Khabarovsk	2	Pavlodar	8
Khamza	2	Perm'	11
Kherson	6	Ryazan'	20
Kirishi	20	Saratov	6
Komsomol'sk	3	Syzran'	12
Krasnodar	2	Tuapse	3
Krasnovodsk	8	Ufa	36
Kremenchug	16	Ukhta	2
Kuybyshev (two plants)	34	Volgograd	10
Lisichansk	15	Yaroslavl'	11

Sources: Sagers (1984, pp.20-21).

European countries of CMEA) (WEFA, 1983). Approximately 7 million tons were imported in 1980, making Soviet net exports about 112 million tons. This amount is allocated among four demand nodes. Exports to Poland, Czechoslovakia, East Germany and Hungary (58.5 million tons) are assumed to be entirely through the "Friendship" Pipeline. These exports are represented by demand nodes located at Uzhgorod and Brest, the pipeline's terminii at the Soviet Border (Figure 4.2). Exports to East Germany and Poland (32.7 million tons) are assigned to Brest, while exports to Czechoslovakia and Hungary (25.8 million tons) are assigned to Uzhgorod. Exports to Bulgaria and Romania (14.2 million tons) are assumed to be shipped by tanker from Novorossysk on the Black Sea, and are represented by an export demand node there. All remaining net exports are allocated to a foreign demand node at Ventspils, the Baltic Sea port (Figure 4.2).

Perhaps as much as 35 percent of these remaining exports are shipped to Mediterranean countries (mainly France and Italy), so they probably originate in the Black Sea ports of Odessa, Novorossysk, and Tuapse rather than Ventspils, the terminal for deliveries to northwest Europe. However, allocating all remaining exports to Ventspils should not significantly affect the final solution.

In addition to these supply and demand nodes, 58 intermediate nodes are added to the network to allow for junctions, changes in capacity, and proper geographic alignment. Nearly half of these are on railroad connections to remote refineries in Central Asia and the Far East that are not served by the main pipeline system (Figure 4.2).

Bounded arcs represent pipelines and other transport links in the system. The upper bound of each arc is determined solely by the size of the pipe, without taking into account possible differences in pumping capacity. The average carrying capacities for the various pipeline sizes are given in Dienes and Shabad (1979, p.63). These range from an annual rate of 8 million tons for a 530 mm pipe to 75 million tons for a 1220 mm pipeline. Pipeline alignments and sizes are taken from several sources, the most important of which are Dienes and Shabad, various "News Notes" in Soviet Geography, Yufina (1978), Shcherbina et al. (1981), and various issues of the journals Neftyanoye khozyaystvo and Stroitel'tsvo truboprovodov.

In addition to pipelines, arcs represent major tanker movements in the Caspian and rail traffic to Central Asia and the Far East. In 1980, about 91 percent of crude moved by pipeline (Biryukov, 1981, p.7). The small amount moving by rail was mostly to supply the small remote refineries in Central Asia and the Far East not on the main pipeline system, although some movements occurred for technological reasons (e.g., to maintain the purity of special crudes) and others because of the lack of pipeline connections. Some crude is moved by tanker, mainly in the Caspian, while river traffic (principally along the Volga) is negligible and ignored (Dienes and Shabad, 1979, p.63).

Thus sea routes in the Caspian between the key points of Baku, Krasnovodsk, Makhachkala, Shevchenko and Astrakhan are included in the abstracted network as are rail links to the refineries in Central Asia and the Far East (Figure 4.2). The refineries at Khamza and Fergana in Uzbekistan are linked to Krasnovodsk in Turkmenia and Pavlodar in Kazakhstan by arcs representing railroads. The refineries at Komsomol'sk and Khabarovsk in the Far East are linked by railroad with Angarsk (Figure 4.2). No data are available to determine transmission capacities for these sea and rail arcs, so the model is allowed to determine the optimal flows of crude in the affected areas unhindered by capacity constraints. This does not significantly affect the overall solution, because the amounts shipped are relatively small. Refinery throughput in Central Asia and the Far East was only about 5 percent of the USSR total in 1980.

Distance is used as a surrogate for transport cost in the model, as there is a close correlation between such costs and distance, especially in pipelines (Tretyakova, 1985). Distances are calculated spherically from the latitude and longitude of the terminal nodes for each arc, making the length of an arc the cost of using it. Pipeline transport costs per ton-km in the USSR are about half the rail costs, and tanker transport costs are 60-70 percent those of rail (Tretyakova, 1985; JPRS, 1984). Because of this differential in cost, distances for rail or tanker shipments are therefore doubled.

D. ANALYSIS OF FLOWS: 1980, 1985

The spatial pattern of model-generated flows for 1980 is shown in Figure 4.3. The pattern is dominated by the flow from the West Siberian oil fields to refineries in the western USSR. This flow occurs along two basic routes. The largest (190 million tons between Surgut and Tobol'sk) is along the older route from Surgut through Tobol'sk to Ufa and Kuybyshev. The magnitude of the flow is approximately at assigned capacity (195 million tons). The second flow of 75 million tons follows the new northern route from Surgut through Perm' to Gor'kiy and Yaroslavl'. This flow is at the rated capacity of the line, implying that its construction was warranted. Thus the movement of petroleum from West Siberia westward is roughly at transmission capacity.

These results indicate that additional pipeline capacity from West Siberia was needed in the 11th Five-Year Plan (1981-85). Initially, the increment in annual output planned for the region was 86 million tons, from 313 million tons in 1980 to 399 million tons in 1985, concomitant with a 1985 production goal for the USSR as a whole of 630 million tons ("News Notes", _SG_-April 1982, p.278). In 1983, for the first time, West Siberia did not fulfil its production target, with a production of 369 million tons, 3 million tons short of the plan ("News Notes", _SG_-April 1984, pp.265-266). West Siberian production in 1984 was 9 million tons below the plan, so production was about 378 million tons ("News Notes", _SG_-April 1985, pp.289-290). Because of these problems in the important West Siberian fields, Soviet oil production peaked in 1983 at 616 million tons. Thus, four years into the 11th Five-Year Plan, the production increment in West Siberia was 65 million tons.

The revised plan for 1985 called for national production to reach 628 million tons of petroleum,

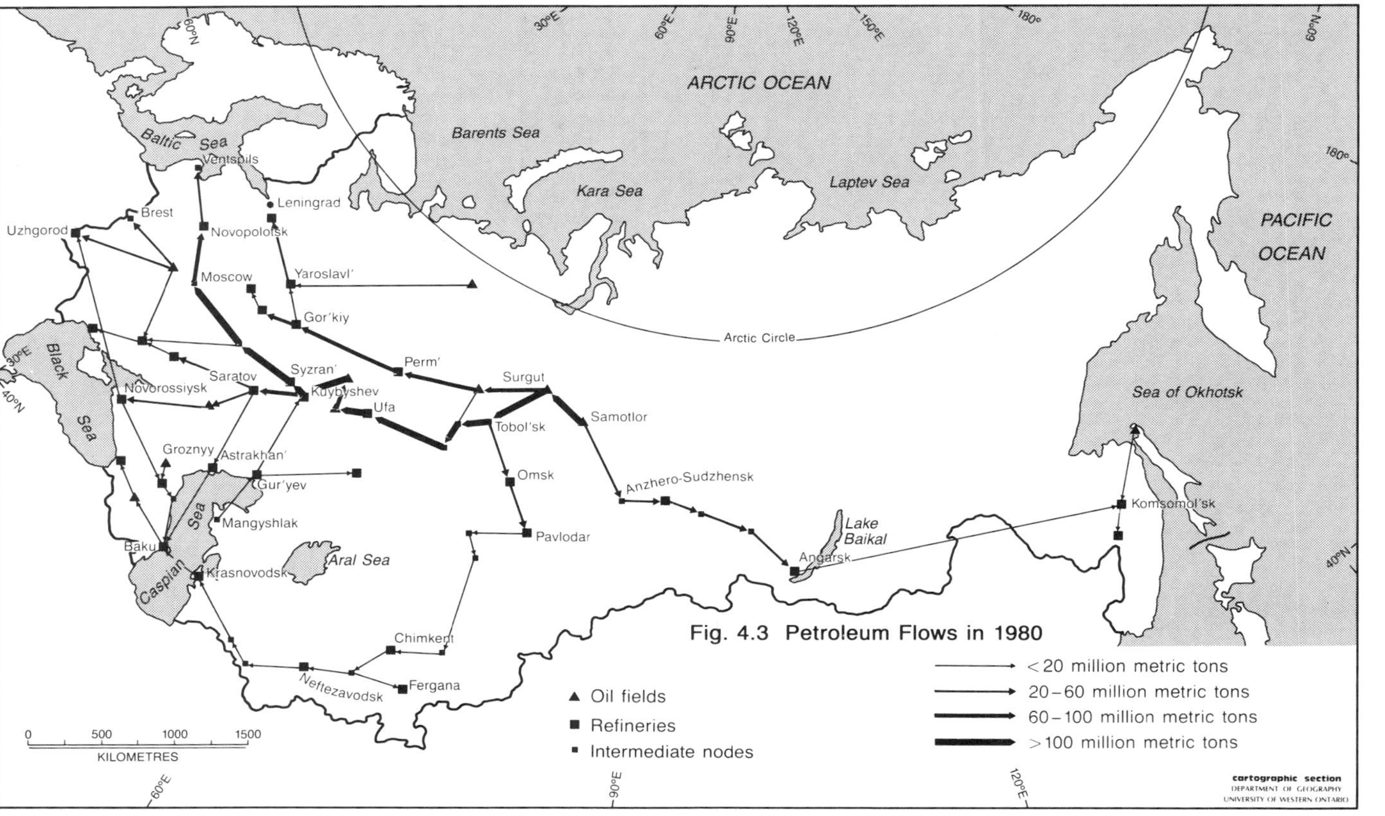

Fig. 4.3 Petroleum Flows in 1980

of which two-thirds was to come from West Siberia (Ekon. gazeta, December, 1984, p.11). The plan for West Siberia was 402 million tons, which was far too ambitious given recent performance; national production slumped to 595 million tons, with the West Siberian component falling slightly to about 376 million tons.

This production increment of about 63 million tons between 1980 and 1985 in West Siberia would appear to justify the construction of another pipeline from West Siberia. The five-year plan originally included the construction of a pipeline from the Urengoy gas field in the northern part of Tyumen' Oblast (for gas condensate) through Noyabr'sk (Kholmogory) and Perm' to Kuybyshev in the Volga region ("News Notes", SG-April 1982, p.282). The line was slated to follow the new northern route through Surgut as far as Perm' before turning south (Figure 4.2). This pipeline has been realigned. It now originates at Noyabr'sk and goes from there to Perm' as originally planned, but then extends from Perm' to Klin, near Moscow. The section between Perm' and Klin was scheduled for completion in 1984, with the entire project to be finished in 1985 ("News Notes", SG-April 1985, p.291). Apparently the Noyabr'sk (Kholmogory) district is considered to hold considerable promise, although production so far has been disappointing and far below expectations ("News Notes", SG-April 1984, pp.265-267).

The construction of the pipeline has been very confusing, partly because of the abruptness of the realignment. For example, the map published in Sotsialiticheskaya industriya (January, 3, 1985, p.1) shows the previous alignment to the Volga region, whereas the map published in Ekonomicheskaya gazeta (January, 1985) shows the alignment to Klin. The destination of Klin is also confusing, because the city does not have a refinery, nor is it known to be on any other oil pipelines.

The flow of Urengoy gas condensate, originally scheduled to be fed into the new pipeline at Kholmogory, has also be reoriented. The pipeline from Urengoy is now an intraregional connection within West Siberia, supplying the new petrochemical center at Tobol'sk and a condensate processing plant in Surgut (Izvestiya, May 4, 1984, p.3; Pravda, June 30, 1983, p.2).

In the set of modeled flows for 1980, the largest single flow in the system is between the oil fields in Kuybyshev Oblast and Kuybyshev refinery complex (282 million tons, or about 47 percent of national production), as the West Siberian flow is joined by production from the Volga-Urals fields (Figure 4.3). This massive flow is just under the assigned upper capacity for the arc (287 million tons).

At Kuybyshev, the main flow from West Siberia (augmented by Volga-Urals production) bifurcates, with a flow of 152 million tons following the "Friendship" Pipeline to the western border at Uzhgorod and Brest, while the other flow of 86 million tons goes south through Saratov to the Ukrainian refineries and the North Caucasus. It is here, rather than farther east that the arcs become capacitated (i.e., flow is equal to or greater than initially assigned capacity). The flow between Kuybyshev and Syzran' required to make the solution feasible is 152 million tons, or 127 percent of the arc's originally assigned capacity, and the Saratov-Volgograd arc is capacitated with a flow of 31 million tons, 24 percent over assigned capacity (Figure 4.3). This indicates that extra capacity is needed at these points for the efficient distribution of crude petroleum. In fact, a short pipeline segment between Saratov and Volgograd was completed in 1985, evidently to alleviate this bottleneck (Summary of World Broadcasts, August 16, 1985, p.7).

Another area requiring more capacity will be the pipeline between Tobol'sk and Omsk. In 1980, this line is modeled as carrying 46 million tons, approximately its assigned capacity of 45 million tons. This flow supplies the demand at Omsk and Pavlodar as well as the Uzbek refineries after crude apparently is transhipped to rail for distribution southward from Pavlodar. When the new refineries at Chimkent and Neftezavodsk and the planned second unit at Pavlodar come onstream (the refineries at Fergana and Khamza in Uzbekistan also will be connected into the main pipeline system), an additional flow of 20 to 24 million tons will be required through the Tobol'sk-Omsk line. Since the line is already apparently fully utilized, carrying capacity between Tobol'sk and Omsk will have to be expanded by a like amount.

It is interesting that the old trans-Siberian line from Omsk to Kurgan is not used in the model, as is the section of the pipeline from Omsk east to Anzhero-Sudzhensk. Demand at the Angarsk refinery is apparently being met by crude moving through the 1220 mm pipeline directly from Samotlor through Anzhero-Sudzhensk, and the new pipelines between West Siberia and the Volga-Urals region have apparently obviated the use of the old trans-Siberian line from Omsk.

Surprisingly, the model indicates that the major transmission difficulties in 1980 existed in the European USSR rather than West Siberia, particularly in the older petroleum-producing areas in the North Caucasus and Azerbaijan. As local production in these areas has dropped since 1970, crude has had to be brought in from the Volga-Urals region and then West Siberia to keep their refineries operating.

All three lines into the area in 1980 are shown as being at their upper bound: Saratov-Volgograd, Lisichansk-Gorlovka and Saratov-Astrakhan (Figure 4.3). Demand at Novorossysk, Tuapse, and Krasnodar is shown as being largely met by West Siberian crude, but supply problems exist further south.

These constraints in the pipeline network force Caspian tanker traffic to occur, despite its high cost, as the model attempts to satisfy refinery demand in the area. The model produces a flow of 4 million tons to Baku across the Caspian Sea from the Turkmen republic and a flow of 8 million tons from Astrakhan (Figure 4.3). Apparently, it is the absence of transshipment costs in the model which allows an unrealistic allocation of Siberian crude through Pavlodar and the Central Asian rail network to Krasnovodsk for further shipment by tanker to Baku (Figure 4.3). The model evidently uses this flow to replace the known movement of Mangyshlak crude to the Baku refineries ("News Notes", SG-October 1983, pp.625-627). The lengths to which the model must go to satisfy demand at Baku illustrates why a shortage of crude has caused the Baku refineries to operate at less than full capacity in recent years ("News Notes", SG-April 1984, p.268).

The fully utilized lines at Lisichansk and Saratov allow only a half a million tons to trickle through Tikhoretsk to Groznyy and Baku, while it is known that a much larger volume of West Siberian crude has to be brought in to supply the Groznyy refineries because of the decline in local production (Sagers, 1984, p.31). Thus, although not entirely accurate because of modeling deficiencies, the flow pattern in the Caucasus illustrates the difficulties engendered in the area by the decline of local production in the last 10 years.

The various transmission difficulties noted above largely confirm the need for three post-1980 developments in the Soviet distribution system. One is the extension of the new northern pipeline from Yaroslavl' to Novopolotsk, accomplished in 1981, the second is an important new line connecting the West Siberian oil fields with Baku through Saratov, Volgograd and Groznyy, which went into operation in 1983 ("News Notes", SG-October 1983, pp.625-627), and the third is the line between Saratov and Volgograd, completed in 1985.

A second network solution, incorporating these arcs, was obtained. The addition of the Yaroslavl'-Novopolotsk arc lowers total system costs by 2.5 percent, a significant improvement in efficiency. This represents a savings of 29.2 billion ton-kilometers in the model, which translates into about 32 million rubles in operating costs per year. The average cost for transporting petroleum by pipeline in 1980 was 1.12 rubles per thousand ton-km (Tretyakova, 1985). This additional arc also allows some of the demand at Novopolotsk, Mazeikiai, and Ventspils to be served from the north, easing the constraint between Kuybyshev and Syzran'. While the Kuybyshev-Syzran' arc was capacitated in the previous run with a flow of 152 million tons, the flow drops below its assigned capacity to 115 million tons with the addition of the Yaroslavl'-Novopolsk arc. Furthermore, the construction of the lines to Baku and between Saratov and Volgograd alleviates considerably the supply problems noted previously for the Caucasus, although the situation here remains tight.

A 1985 scenario, incorporating new pipeline capacity, additions to refineries and production location changes, was developed to evaluate the impact of the new Kholmogory-Klin line and the further concentration of petroleum production in West Siberia. In the scenario, national production was assumed to equal 600 million tons, with 65 percent originating in West Siberia and 22 percent in the Volga-Ural fields. New refinery capacity of 6 million tons each was added at Lisichansk, Baku, Mazeikiai, Achinsk, Chimkent and Neftezavodsk, and 3 million tons at Tobol'sk. Because of the lack of evidence of construction activity at Pavlodar, its planned second unit is not included. Exports in the 1985 scenario were set at 125 million tons, 68 million tons of which are to CMEA countries. This is in accordance with the recent trend for Soviet oil exports to increase despite slowing production, with the bulk of the increase going to Western countries, but the situation has become extremely changeable in recent years.

Three main constraints are revealed by the 1985 flow model. Two result from increased exports, with existing pipeline capacity becoming constrainted at the petroleum exporting ports of Novorossysk and Ventspils. The third constraint occurs in the Tobol'sk-Omsk-Pavlodar line as expected (see above) because of the additional refinery capacity at Chimkent and Neftezvodsk. Thus, a project that should be in the 12th Five-Year plan would be a second pipeline between West Siberia and Pavlodar.

Only portions of the new Kholmogory-Klin line are used in the 1985 solution. The arcs between Surgut and Perm' and between Perm' and Gor'kiy are fully used, whereas the projected flows on the other arcs that comprise the pipeline are less than the existing (lower) capacities of 1980. In fact, the final section of the line, from Yaroslavl' to Moscow (Klin) carries no flow at all. Thus, given our limited knowledge of the system, it appears that a policy of upgrading selected sections of the existing Surgut-Polotsk pipeline (i.e., construction of a second string between Surgut and Gor'kiy) would have been better than the construction of an entire new line, particularly extending to Klin.

E. CONCLUSIONS

Except for some flows in the Caucasus and Caspian Sea, the hypothetical flow patterns generated by the model generally agree with what little information exists on actual Soviet petroleum flows. Total transport costs for the model in billion ton-kilometers are 90 percent of the actual total for the USSR in 1980 (1128.3 versus 1256.9 billion ton-kilometers) (Tretyakova, 1985). Since the model appears to adequately simulate the Soviet system, this implies that overall, the Soviet pipeline network seems to be operating relatively efficiently.

Several interesting conclusions stem from the analysis. First, the main transport constraints in the system in 1980 were in the European USSR, in the older oil-producing areas, rather than in the east at the origin of most of the USSR's petroleum. Second, this problem evidently has been alleviated to a considerable extent by the construction of additional pipelines in the region, so the model correctly anticipates subsequent developments. Third, the new line constructed from Surgut to Novopolotsk greatly increased the efficiency of petroleum transmission, supporting the necessity of its construction. Fourth, the model shows the pipelines from West Siberia to the European USSR to be operating close to capacity in 1980, demonstrating the reason for new pipeline construction from Noyabr'sk to Klin, although a policy of selective upgrading of the existing line appears more effective.

V
The Transportation of Refined Petroleum Products

A. INTRODUCTION

Recently, the railroads, the mode which has traditionally borne most of the USSR's transport burden, have begun to falter from an increasing strain (Hunter and Kaple, 1982; 1983). One of the most important elements in this escalating transport burden has been the recent changes in fuel supplies and the evolution of regional fuel mixes (Dienes, 1983b, p.405), which for the railroads involves the transport of coal and petroleum products. Together, they account for nearly 30 percent of all railroad freight movements (Narkhoz SSSR 1984, p.338).

Although the commodity creating the heaviest burden for the railroads is undeniably coal (see chapter VI; Dienes, 1983b, pp.398-401; Hunter and Kaple, 1983, p.43), petroleum adds considerably to the overall strain. Until 1981, petroleum freight was second in importance to coal in generating rail traffic. While recently it has slipped to third place behind building materials, petroleum freight still accounted to 12.3 percent of all railroad ton-kilometers and 11.0 percent of all tons originated in 1984 (Narkhoz SSSR 1984, p.338).

The bulk of the petroleum freight carried by the railroads is comprised of refined products, as over 90 percent of crude is shipped by pipeline (Biryukov, 1981, p.7). Of the 422.7 million tons of petroleum freight carried by the railroads in 1980 (Narkhoz SSSR 1984, p.338), about 365 million tons were refined products. Their movement generated approximately 364 billion ton-kilometers of freight turnover, or about 10.6 percent of the USSR total. In 1980, petroleum freight turnover on Soviet railroads was 460.8 billion ton-kilometers (Narkhoz SSSR

1980, p.295). This includes 96.5 billion ton-kilometers incurred in carrying crude petroleum rather than refined products (Tretyakova, 1985). The difference, 364.3 billion ton-kilometers, is the freight turnover for refined products. The tonnage carried can be estimated from this by dividing it by the average length of haul or by applying the percentage of rail in all shipments of refined products; both methods yield nearly the same amount. This is a fairly significant amount, as it exceeds the freight movement of most of the other major commodity groups (Narkhoz SSSR 1980, p.295).

This chapter examines the problem of transporting petroleum products in the USSR. However, unlike the transport problem posed by the increasing volume of coal movements, which is largely the result of the distribution of the resource base (see chapter VI; "News Notes", SG-December 1982, pp.769-772), problems in the transportation of refined products largely result from the geographic distribution and capacity of existing refineries and the mix of the refined products produced (Office of Technology Assessment, 1981, p.64).

This is because the refining industry is heavily concentrated in relatively few areas -- the Volga, North Caucasus, and Azerbaijan (the traditional oil-producing regions). They therefore have a chronic surplus of refined products, while refining capacity and production remain in relatively short supply in many other consuming areas (Wilson, 1983, pp.178-181). Thus, part of the strain on the transport system is a consequence of the resulting interregional movement of millions of tons of refined products.

This inefficient locational pattern reflects the previous raw materials orientation of the refining industry. Before the development of an interregional pipeline system, the relatively small amount of crude oil produced was refined mainly at the production sites (Dienes and Shabad, 1979, p.64). Thus, the Caucasus and Volga regions accounted for 85 percent of refinery throughput in the mid-1950s (Gankin *et al.*, 1972, p.102). But with the growing importance of petroleum in the economy and the development of pipeline transport for crude, it became more efficient to locate new refineries in market areas (Dienes and Shabad, 1979, p.64; Sagers, 1984, p.3). This is because the distribution of refined products by rail is 2 to 3 times more expensive per ton-kilometer than crude shipped by pipeline (Tretyakova, 1985). As a result, the refining industry experienced considerable dispersion after 1960, as several new plants were built in the major consuming regions. Despite this, the location of refining capacity still reflects the earlier development, with a heavy concentration in the older refining regions (Sagers, 1984, p.39).

Another aspect of this problem of over-concentration is the large size of Soviet refineries. The USSR had only 48 refineries in 1980, which refined 441 million tons of crude (Sagers, 1984, pp.20-21). Thus the average plant size was 9 million tons; most are much larger. In contrast, U.S. refining capacity is far more dispersed, as it is comprised of a much larger number of smaller plants. The U.S. had 301 operable refineries in 1982, which refined 11.8 million barrels per day, or about 590 million tons (Energy Information Administration, 1983, p.5). This averages out to only about 2 million tons per plant.

A separate facet of the transport burden is the considerable volume of cross-hauls and other "irrational" flows of refined products. This is largely the result of discrepancies between the output mix of refineries and local consumption patterns, although "irrational" shipments also result from products not always being obtained from the nearest source of supply. The magnitude of these "unnecessary" shipments is unknown, although anecdotal evidence in the press suggests that the problem is quite significant (e.g., Kontorovich, 1982, p.56; Gankin _et al._, 1972, pp.113-119; "News Notes", _SG_-September 1969, p.423; December 1976, p.714). An estimate of the magnitude of these flows will be one of the results of the network analysis presented below.

The transportation of refined products is a very difficult area to evaluate because of extremely limited information on oil refining. The data needed (refinery size, regional consumption levels, and transport capacity) are not readily available; a variety of estimates and approximations must be used. There is also virtually no information on the output mix of individual refineries. In fact, only rough estimates are available on the production of various refined products for the USSR as a whole (see Campbell, 1976, p.46; U.N., 1983, pp.556, 570, 579; Sagers and Tretyakova, 1985, p.8). This severely limits the scope of analysis.

However, the limited data available have been assembled to allow the Soviet distribution system of 1980 to be modeled. The generated set of optimal, hypothetical flows for the system are used to make some inferences about the actual magnitude and direction of refined product movements, as well as the efficiency of the location of refining facilities. Initially though, what is known about actual movements of refined products in the USSR is presented.

B. REFINED PRODUCT MOVEMENTS

1. Modal Distribution

The heavy burden borne by the railroads in transporting refined products in the Soviet Union is unique among the industrialized countries (Dienes, 1983b, p.398). This is because the enormous task of distributing refined products throughout the huge area of the USSR is shouldered almost entirely by the railroads. They accounted for 79.2 percent of the primary shipments of petroleum products (i.e., those from the refineries) in 1981 (Table 5.1).

The transport of refined products by river and sea tankers is fairly minor, accounting for only 6.8 and 2.7 percent of shipments in 1981, respectively (Table 5.1). Soviet waterways carry such a small proportion mainly because of their poor location relative to the main centers of industry and the seasonal limitations imposed by the cold climate (Perova and Maslov, 1984, p.10; Hunter, 1968, pp.10-13).

TABLE 5.1

Primary Transportation of Petroleum Products in 1981*

Mode	Shipments (pct)	Freight Turnover (pct)	Average Length of Haul (km)
Railroad	79.2	79.6	973
Pipeline	11.3	8.3	706
River tanker	6.8	8.7	1236
Sea-going tanker	2.7	3.4	1200
All modes	100.0	100.0	967

*These data refer only to shipments within the country.

Source: Grigor'yev *et al*. (1984, p.61).

Of course, one reason for the large proportion carried by rail is the widely scattered distribution of destinations typical for refined products. However, rail even dominates in areas of bulk movements (Dienes and Shabad, 1979, p.63), usually the province of product pipelines. This is because product pipelines, the least costly mode of transport (see below), remain underdeveloped (see Figure 5.1). Of the total oil pipeline network of 69,700 kilometers in 1980 (Narkhoz SSSR 1980, p.305), only 11,300 kilometers were for refined products (Vlasov, 1984, p.21). Thus only 10 to 12 percent of refined products moved by pipeline in 1980 (Perova and Maslov, 1984, p.10; Biryukov, 1981, p.7), or about 55 million tons, which generated approximately 36 billion ton-kilometers of freight turnover. This represents the 1196.8 billion ton-kilometers on all petroleum pipelines (Narkhoz SSSR 1980, p.305) less the 1160.4 billion ton-kilometers for crude petroleum pipelines (Tretyakova, 1985).

However, the situation for product pipelines has been improving, relieving some of the congestion on the Soviet railroads. In 1975, pipelines carried only 39.5 million tons, or 9.4 percent of primary shipments (Maslov and Perova, 1976, p.30). The tonnage carried by product pipelines increased by 40 percent between 1976 and 1980 as the network expanded by two thousand kilometers (Biryukov *et al*., 1982, p.31). A vigorous development program also was scheduled for the 11th Five-Year Plan (1981-1985). The construction of 6,500 kilometers of new pipeline for petroleum products was planned, to increase the tonnage carried by about 30 percent ("News Notes", *SG*-April 1982, p.282). During the first three years of the 11th Five-Year Plan, approximately 3000 kilometers of product pipeline were commissioned (Perova and Maslov, 1984, p.7), allowing the volume of shipments to increase 63 percent between 1975 and 1982 (Grigor'yev *et al*., 1984, p.60). This implies that about 64 million tons of refined products were shipped by pipeline in 1982.

In addition to their relative underdevelopment, another reason for the limited contribution of pipelines in transporting refined products is the Soviet output mix. Since 1970, residual fuel oil (mazut)

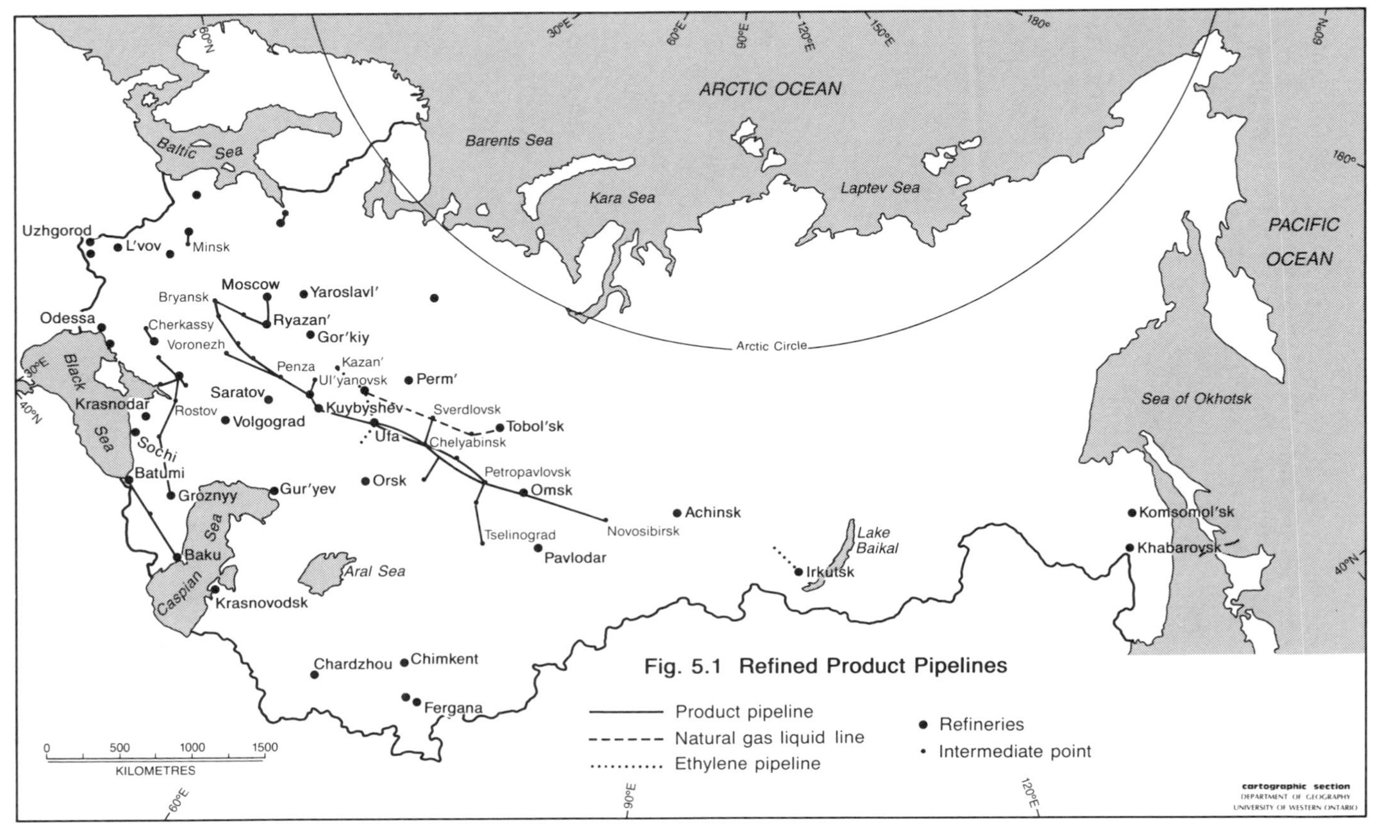

Fig. 5.1 Refined Product Pipelines

rather consistently has accounted for about 38 percent of Soviet production, making it by far the dominant product in the output mix (Sagers and Tretyakova, 1985, p.9). Gasoline and diesel fuel together account for only about 41 percent of Soviet output. In comparison, only 15 percent of U.S. refined products is composed of residual fuel oil, while 56 percent is gasoline and diesel fuel (U.N., 1981, pp.571, 556, and 585). Since mostly light products (primarily gasoline and diesel fuel) are shipped by pipeline (mazut is generally too heavy), a major factor limiting the distribution of refined products by pipeline in the USSR is the relatively small share of motor fuels in Soviet output (Perova and Maslov, 1984, p.10). However, the amount shipped by pipeline still is far less than the quantity of gasoline and diesel fuel produced.

2. Length of Haul

In the mid-1970s, the refined product which accounted for the greatest volume of interregional shipments (nearly 80 percent) was mazut (Danilov, 1977, p.153). This probably is due to its dominance in the overall output mix. Also, its average length of haul in 1975 was relatively high; of the major products, only kerosene had a longer average haul (Table 5.2). However, mazut experienced the greatest relative decline in its average length of haul between 1975 and 1982, from 1084 kilometers in 1975 to 906 in 1982, giving it one of the lowest among the major products (Table 5.2). This reflects the dispersion of the oil refining industry during this period; since mazut is the predominant product of Soviet refineries, this would make it more of a ubiquity than any other refined product. Thus, while it still may be the major product in interregional movements because of its heavy weight in the output mix, it is probably no longer as dominant as it was in the mid-1970s.

Like mazut, kerosene (particularly for lighting and household use) had a significant role in interregional shipments in the mid-1970s (Danilov, 1977, p.153). This is probably because kerosene is shipped the farthest of any of the major refined products (Table 5.2). Also, unlike the other refined products it has experienced no decline in

TABLE 5.2

Average Length of Haul for Specific Products
(kilometers)

Product	1975	1976	1977	1978	1979	1980	1981	1982
All products	1105	1088	1074	1055	1017	997	967	965
Diesel Fuel	1084	1063	1054	1057	984	979	926	917
Gasoline	982	973	966	971	838	910	877	875
Mazut	1084	1067	1042	990	957	925	882	906
Kerosene	1216	1187	1176	1173	1203	1217	1203	1223

Note: For kerosene, the average haul in 1980 is calculated from a weighted average for all products using the shares of the various products in the Soviet output mix as weights. In this calculation, the weight for kerosene also includes a number of minor products. The shares are from Sagers and Tretyakova, (1985, p.8). The average length of haul for all other years in the table are calculated from indexes given in Grigor'yev *et al*. (1984, p.61).

Source: Grigor'yev *et al*. (1984, p.61).

its average length of haul since 1975 (Table 5.2). Thus, its role in interregional shipments undoubtedly has increased. This may reflect the fact that its main use is in aviation, a far-flung consumer.

In contrast, gasoline has the shortest length of haul, and by a considerable margin (Table 5.2). This indicates that it plays a fairly minor role in interregional shipments compared with the other products. However, its length of haul in 1982, 875 kilometers, is still a considerable distance, indicating that a great deal of gasoline is shipped between regions.

3. Flow Patterns

Relatively little is known about the geography of refined product movements in the USSR. Although several general descriptions of the main movements of refined products are available (Kazanskiy, 1980, p.41; Danilov, 1977, pp.153, 157-159), nothing is known about the magnitudes of these flows. Unfortunately, these sources also refer to the situation that existed in the mid-to-late 1970s. The general pattern of the flows of refined products as given in these sources is summarized below and shown in Figure 5.2.

Shipments of refined products from the Volga-Urals refineries (e.g., Kuybyshev, Ufa, Salavat, Ishimbay, Syzran') through Penza to the Moscow area and Khar'kov, and through Volgograd to Novorossysk, are said to dominate the pattern of distribution (Figure 5.2). Much of this movement is handled by rail, although the Kuybyshev-Bryansk pipeline (Figure 5.1) carries some of the light products involved. Some of the shipments to the Central region are by tanker up the Volga River to ports such as Gor'kiy, Kostroma, Yaroslavl', and Kalinin (Figure 5.2), where they are transferred to rail for final distribution.

There is also a major flow of refined products east from the Volga refineries (particularly Ufa) through Chelyabinsk and Omsk, supplying the southern Urals, West Siberia (Novosibirsk and the Kuzbas), and northern Kazakhstan (Figure 5.2). Much of this

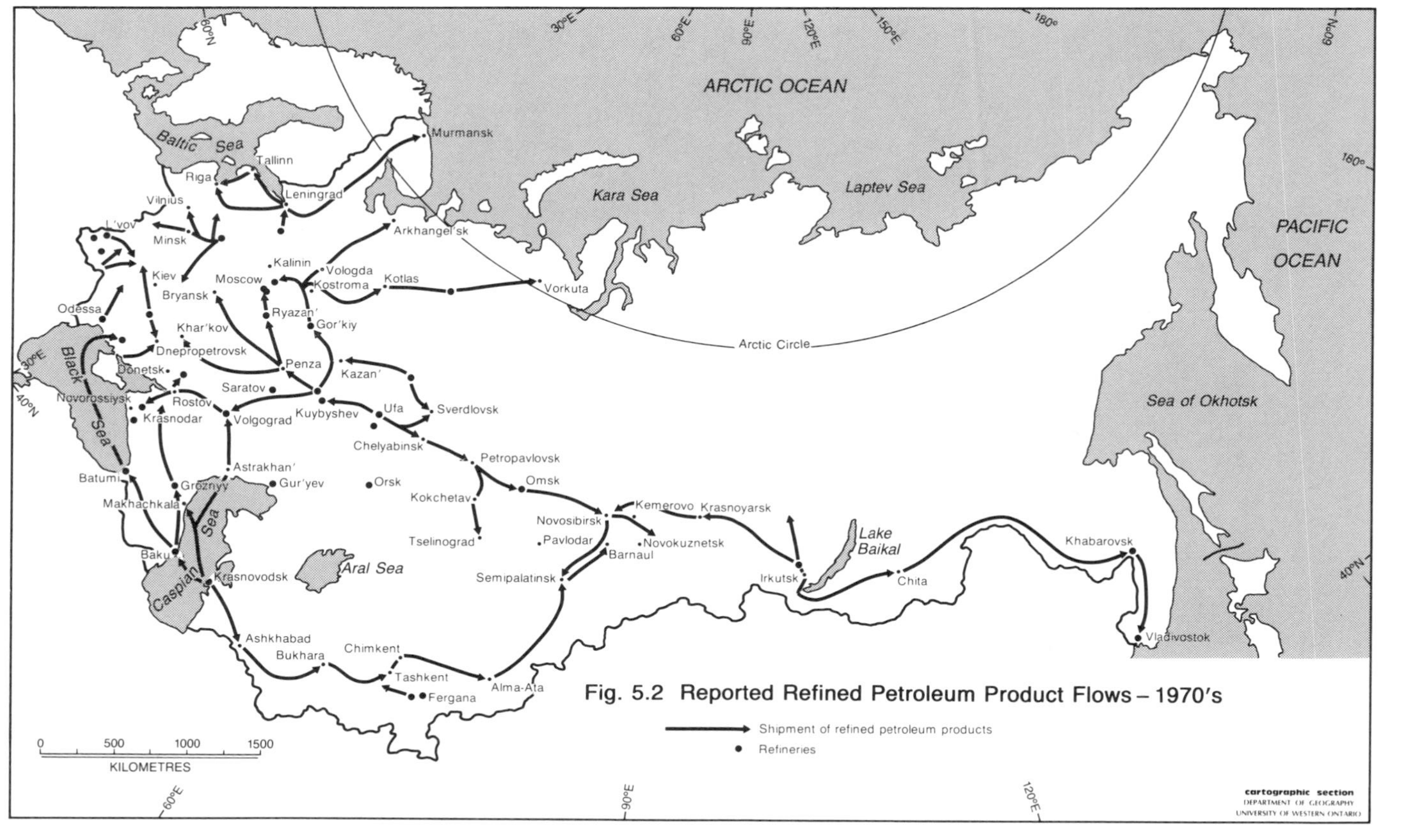

Fig. 5.2 Reported Refined Petroleum Product Flows – 1970's

is comprised of light products shipped by pipeline (Figure 5.1).

The Angarsk refinery supplies mostly East Siberia. However, some of Angarsk's output is shipped to the Far East ("News Notes", SG-November 1977, p.702), where production is less than consumption, and some products are shipped west from Angarsk through Krasnoyarsk to Novosibirsk and Eastern Kazakhstan (Figure 5.2).

About 80 percent of the small Central Asian output is used within the region. However, the intraregional flows do involve a long haul from Krasnovodsk through Ashkhabad to Tashkent (Figure 5.2). Most of the remainder of regional production is shipped to Siberia and the Far East through Alma-Ata and Barnaul, although some is shipped across the Caspian Sea to Baku and Makhachkala, and from there to the Ukraine (Figure 5.2).

Similarly, much of the output of the large refining complexes at Baku and Groznyy in the Caucasus is shipped north to the Ukraine and the Central region, mostly by rail, although some goes by pipeline (Groznyy-Armavir-Donbas), and some by tanker across the Black Sea or Caspian/Volga (Figure 5.2).

The Ukrainian refineries are not large enough to satisfy regional consumption needs, and the same is true for the central refineries around Moscow; thus their output is consumed locally. However, the Yaroslavl' refinery supplies a large area extending north from Yaroslavl' through Vologda, Arkhangel'sk, and Vorkuta, while the Kirishi refinery supplies the Leningrad and Murmansk areas, and some of the Baltic region (Figure 5.2). The Polotsk refinery supplies Belorussia and the remaining needs of the Baltic region.

This general pattern has changed somewhat since the 1970s. New refineries have been built in Belorussia (Mozyr'), Kazakhstan (Pavlodar), the Baltic (Mazeikiai), the Ukraine (Lisichansk), and the Volga region (Nizhnekamsk) (Sagers, 1984). Also, there has been a considerable increase in gas use, particularly in the Urals and Central regions (see Chapter III), perhaps replacing some petroleum consumption.

Changes in the supply pattern described above are evident from a few clues in recent sources. Biryukov et al. (1982, p.30), apparently referring to 1980, report a surplus of production not only for the Urals-Volga region, but also for the Central and Volga-Vyatka regions, with a deficit in the Ukraine, Moldavia, Kazakhstan, Central Asia, the Baltic, and the Far East, giving rise to a flow of tens of millions of tons of petroleum products between them. By the early 1980s, apparently a surplus also had developed in Belorussia and the Northwest (Grigor'yev et al., 1984, p.60).

Refinery expansion in the 11th Five-Year Plan (i.e., additional units at Mazeikiai, Lisichansk, and Pavlodar, and new refineries at Achinsk, Chimkent, and Neftezavodsk) (see Sagers, 1984) is designed to reduce transportation costs by meeting the needs of these deficit areas (Biryukov et al., 1982, p.30). Bringing refinery capacity to consuming regions is identified as an important "reserve" in rationalizing freight movements. This reflects the general Soviet penchant for minimizing overall transport effort (see Houston, 1969). The success of Soviet planning in achieving this goal will be evaluated in the network analysis below.

C. THE PRODUCTS DISTRIBUTION SYSTEM

1. The Problem of Heterogeneous Products

The spatial dimension of refined product flows for 1980 is analyzed. This network consists of supply nodes (oil refineries), demand nodes (major cities, mazut-burning electrical power stations, and export points), and bounded arcs (railroad lines, product pipelines, and some tanker routes).

The major shortcoming in the analysis of this particular problem is it assumes a homogeneous commodity is being distributed. However, refined products are very heterogeneous and utilized for a wide variety of purposes. Also, petroleum refining involves several different processes which have a distinct impact upon the product mix (see Sagers and Tretyakova, 1985), and not all Soviet refineries utilize the same mix of processes (see Sagers, 1984). The different kinds of crude utilized at the various refineries also affect the output mix.

If data were available for each refinery on the production of various products and the processes utilized, then a multi-commodity, multi-process model such as that used by ZumBrunnen and Osleeb (1986) to investigate the Soviet iron and steel industry would be appropriate. Unfortunately, these data are not available. Thus, our investigation is forced to use the assumption of a homogeneous commodity to examine what the magnitude of flows would be in an optimal situation in which production at each refinery exactly matched the demand for various products in the refinery's market area. Thus the network analysis can indicate to what extent this ideal does not occur, resulting in cross-hauls between regions of different products. It can also provide some indication of the rationality of the planned refinery expansions.

Differing production costs among the refineries, due to factors such as economies of scale, procurement costs for crude, and the technological level of equipment, are not considered in the allocation model. Differences in production costs are not as important in the USSR as they would be in a market economy, for the simple reason that excess refinery capacity does not exist. Until the early 1980s, all refining capacity was fully utilized regardless of condition (Oil and Gas Journal, 1984, p.25; Sagers, 1984). Thus, relative transportation costs (and the overall volume of output) largely determine a refinery's market area, not differences in production costs. This is because the option of obtaining more output from a cheaper producer is not available.

Secondly, because the cost of distributing crude by pipeline is so much less than distributing refined products by rail, crude is treated as a ubiquity at all refinery locations. This simplifying assumption is used because refinery expansion in the USSR is determined largely by marketing considerations rather than procurement costs for crude

(Dienes and Shabad, 1979, p.64). An exception to this is the Far Eastern refineries at Khabarovsk and Komsomol'sk. They are not connected into the main pipeline network. About half of their throughput is comprised of West Siberian crude which must be railed from the terminus of the main pipeline system at Angarsk.

2. Nodes

Supply nodes in the network are the Soviet refineries or refinery complexes if more than one plant exists in a city. In 1980, there were 42 of these refinery locations. The supply availability at each is determined by their annual primary crude throughput (Table 5.4), although there is actually a loss during refining that varies from plant to plant, but averaged 1.25 percent in the USSR in 1980 (Ekon. gazeta, November, 1981, p.2). Thus the U.N. estimate of total refinery output in 1980 (436.5 million tons) is in agreement with the estimate of aggregate refinery throughput of crude of 441 million tons by Sagers (1984, p.10) after subtracting refinery losses. However, petroleum products are also derived from the refining of gas condensate as well as crude petroleum, so the total output of refined products is higher (Sagers and Tretyakova, 1985, p.8). The available supply at the refineries (441 million tons) is not differentiated by product.

Demand nodes in the system consist of export points, major mazut-fired power stations, and large cities. The latter two represent domestic consumption of refined products. The only attempt to disaggregate consumption by product is for mazut, the principal output of Soviet refineries. Thus, discrepancies that exist between the output mix of refineries and local consumption of various products are ignored. However, an approximation of the relative magnitude of this discrepancy through the cross-hauls of products it engenders will be offered later.

Net exports of petroleum products, estimated to be about 42 million tons in 1980 ("News Notes", SG-April 1982, p.183; Wharton Econometrics Forecasting Associates, 1983; U.N., 1983), are assigned equally to export points at Brest, Uzhgorod, L'vov, Grodno, Odessa, and Ventspils (i.e., 7 million tons each).

TABLE 5.3

Supply Nodes for Petroleum Products

Refinery/City	1980 Throughput (million tons)	Refinery/City	1980 Throughput (million tons)
Achinsk	0	L'vov	1
Angarsk	23	Mazeikiai	7
Baku (three plants)	25	Moscow	10
Batumi	5	Mozyr'	12
Chimkent	0	Nadvornaya	2
Drogobych (two plants)	2	Neftezavodsk	0
Fergana	7	Nizhnekamsk	8
Gor'kiy	25	Novopolotsk	23
Groznyy	15	Odessa	3
Gur'yev	6	Omsk	26
Ishimbay-Salavat (two)	10	Orsk	4
Khabarovsk	2	Pavlodar	8
Khamza	2	Perm'	11
Kherson	6	Ryazan'	20
Kirishi	20	Saratov	6
Komsomol'sk	3	Syzran'	12
Krasnodar	2	Tuapse	3
Krasnovodsk	8	Ufa	36
Kremenchug	16	Ukhta	2
Kuybyshev (two plants)	34	Volgograd	10
Lisichansk	15	Yaroslavl'	11

Source: Sagers (1984, pp.20-21).

Most of the mazut produced is used as fuel in electrical power stations; nearly 115 million tons of mazut were used in 1980 by the Ministry of Electrical Power in generating electricity (Nekrasov and Troitskiy, 1981, p.230). The approximate regional distribution of mazut consumption by central electrical power stations can be estimated from data given in Nekrasov and Troitskiy (1981, pp.232, 224, 225, 228) and Chapter III. Mazut consumption by electrical stations in 1980 was approximately 4.1 million tons in the Urals, 26.0 million tons in the Volga region, 1.4 million tons in West Siberia, 0.3 million tons in East Siberia, 4.7 million tons in Kazakhstan and Central Asia, 76.4 million tons in the other European regions, and 1.6 million tons in the Far East.

These regional totals are allocated among the major mazut-burning stations (all known stations over 300 MW) within each region according to installed generating capacity. Only half of the capacity of stations using mazut for part of the year ("buferniy" stations) is used in determining the allocation. Since there are no major mazut-burning stations in the Far East, the regional total (1.6 million tons) is allocated equally among the three major cities (Vladivostok, Khabarovsk, and Komsomol'sk) to reflect consumption by local heat and power plants (TETs). This amount for each station is well in excess of actual consumption because each also is assigned a share of the amount consumed throughout the region at a number of smaller TETs (heat and power plants). For example, the Karmanovo station (1800 MW) is assigned 8.9 million tons, although its actual annual consumption is only about 2.6 million tons ("News Notes", SG-September 1984, p.547). A total of 37 stations are included as demand nodes in the network.

No data are available on the regional distribution of the rest of domestic consumption of refined products. However, the distribution of all mazut consumption ca. 1978 can be estimated from data in Davydov and Artyukhova (1979). The regional distribution of non-power station consumption is crudely approximated by using large cities to represent consumption sites other than electrical stations. Because of the congruence with the estimates of refinery throughput used here, the U.N. estimate of total refinery output of 436.5 million tons is used, rather than the higher estimate of Sagers and Tretyakova (1985). Thus, a total of 280 million tons, representing the remainder of apparent domestic consumption of refined products (i.e., total refinery output less net exports (42 million tons) less consumption by electrical power stations (115 million tons)), is allocated among the cities of the USSR according to population size (Narkhoz SSSR 1979, pp.18-28); no other data are available for this purpose. City size is used as it is known to reflect the relative industrial importance of the city as well as the size of its housing and communal economy (Huzinec, 1978), two of the largest consuming sectors of petroleum products in the USSR. In 1972, they accounted for 41.1 and 11.0 percent of domestic consumption, respectively (Kurtzweg and Tretyakova, 1982, p.366).

Unfortunately, an allocation based upon city size does not necessarily directly reflect regional consumption patterns by two other major consuming sectors, agriculture and transportation. In 1972, they each accounted for about 15 percent of domestic consumption (Kurtzweg and Tretyakova, 1982, p.366). However, the spatial distributions of agriculture and transportation generally correspond to that of the population (and therefore the urban population used here) (see Lydolph, 1979, pp.236, 242, 447). Thus, perhaps their consumption pattern is captured in at least a gross manner by this method of allocation.

A total of 270 nodes are included, comprising all cities over 100,000 in 1980 as well as all remaining oblast centers. This is a far greater number of cities than previous studies have used to allocate regional electricity and gas consumption (Sagers and Green, 1982; see Chapter III) in order to allow for the much greater dispersion of refined product consumption than the other major forms of energy.

The allocation for Moscow includes the population of neighboring cities over 100,000 (e.g., Balashikha, Zagorsk, Zelenograd, Mystishchi) as does that for Leningrad and Kuybyshev. Not surprisingly, the largest demand node is Moscow, consuming an estimated 27.2 million tons of refined products, in addition to the 5.9 million tons of mazut assigned its power stations. Leningrad, the second largest, has a demand only about half this (13.6 million tons). The other nodes are much smaller. For example, Khar'kov's allocation is 4.1 million tons, with allocations for many of the smaller cities at 0.2 to 0.3 million tons.

These data are aggregated by economic region/republic and shown as a percentage of the national total in Table 5.3. According to these data, the two largest consuming regions are the Center and the Ukraine, followed by the Volga region. These data are further aggregated into four larger regional groupings (Ural-Volga, Caucasus, Other European, and Asian portion) in order to compare them with available 1969 data (Table 5.3). This comparison suggests that these rough consumption estimates for 1980 at least appear reasonable.

In terms of the regional pattern of deficits and surpluses, these consumption estimates, when compared with refinery throughput (Table 5.4), yield results generally concurring with those of Biryukov *et al.* (1982, p.30). According to our regional consumption/production estimates, a massive surplus of refined products exists in the Volga-Urals (53 million tons). The Caucasus and Belorussia also have a surplus (6 and 12 million tons, respectively), while shortages exist in the Ukraine (26 million tons), East Siberia (2 million tons), the Baltic (7 million tons), Central Asia (3 million tons), and Kazakhstan (7 million tons). The major discrepancy between these estimates and the information in Biryukov *et al*. (1982, p.30) is the situation in the Central and Volga-Vyatka regions. Our consumption/production estimates indicate that the region is short about 15 million tons of products whereas the Soviet source indicates that the area actually had a surplus of production. This may result from the greater use of gas in the region to meet fuel needs, which would not be reflected in our estimates of consumption based only on population size.

TABLE 5.4

Apparent Consumption of Refined Products
(percent)

Region	1969	1980
USSR	100	100.0
Ural-Volga	16	18.7
Volga	--	12.6
Urals	--	6.1
Caucasus	14	10.1
Transcaucasus	--	4.7
North Caucasus	--	5.4
Other European	50	52.9
Northwest	--	7.5
Center	--	16.6
Volga-Vyatka	--	2.1
Central Chernozem	--	1.7
Ukraine	--	16.2
Belorussia	--	5.2
Baltic	--	3.0
Moldavia	--	0.6
Asian portion	20	18.3
West Siberia	--	4.7
East Siberia	--	2.3
Far East	--	2.1
Kazakhstan	--	4.7
Central Asia	--	4.5

Sources: 1969 data from Gankin et al. (1972, p.104). 1980 data aggregated from city allocations and electrical station allocations.

In addition to these supply and demand nodes, 95 intermediate nodes are added to the network to allow for junctions and proper geographic alignment in the transportation system. All the intermediate nodes are cities or towns.

3. Arcs

Bounded arcs represent rail lines, products pipelines, and some tanker routes. Automobile transport is ignored since it is mostly used for local (secondary) distribution rather than primary movements from the refineries (Vlasov, 1984, p.23). Most of the arcs in the network represent rail lines, since the railroad is the major mode used. The alignment of the rail lines is taken from several Soviet maps and atlases showing the railroad network (e.g., *Atlas zheleznykh dorog*, 1984).

All known products pipelines from the refineries are included. Their alignments are listed in Sagers (1984) and shown in Figure 5.1. Since several would duplicate arcs already in the network representing rail lines, the redundant (rail) arcs are deleted from the model. We limit our model to one arc per route.

Arcs representing major tanker movements are added, mainly for the heavily utilized Caspian routes. Nearly 67 percent of all (domestic) sea shipments of petroleum products occur in the Caspian (Grigor'yev *et al.*, 1984, p.61). Some tanker routes are added for cities in the Far East and East Siberia not served by rail (e.g., between Vladivostok and Magadan, Petropavlovsk-Kamchatskiy, and Yuzhno-Sakhalinsk). Since 78 percent of all shipments by river tanker are in the Volga-Kama basin (Grigor'yev *et al.*, 1984, p.61), and 60 percent of all river shipments originate at only three ports (Kuybyshev, Ufa, and Kambarka), these critical river arcs are added to the network. Again, redundant (rail) arcs are deleted from the model.

Each of the products pipelines carries less than 10 million tons per year, and some as little as one million tons because of their relatively small diameters (e.g., Ryazan'-Orel, 3 million tons; Lisichansk-Donetsk, 1 million tons). However, no data are available to determine annual transmission capacities for the other modes. As a result, the model is allowed to determine the optimal flows practically unhindered by capacity constraints on the arcs. A limit of 40 million tons is assigned to all arcs, an amount set to exceed the output of any single refinery. But only a few arcs require more than 20 million tons of capacity in the solution, and none require more than 30 million tons.

Distance is used as a surrogate for transportation cost in the model, as there is a close correlation between such costs and distance for rail and for pipelines (Sagers and Green, 1985b; Tretyakova, 1985). Thus no attempt is made to differentiate costs between the various transport modes because rail accounts for the bulk of shipments. However, pipeline transport costs per ton-kilometer for petroleum products are less than half those of rail, and tanker costs are intermediate between the two. The average shipping cost by rail for refined products per ten ton-kilometers is about 2.5 kopecks (Shafirkin, 1978, pp.222-223), whereas that for pipeline shipments is about .95 kopecks (V.G. Dubinskiy, 1977, p.15). Although no specific data are available on the shipping cost of refined products by tanker, the average cost per ten ton-kilometers in 1970 was 2.45 kopecks for all internal water shipments (Shul'ga, 1974, p.9). Since the average cost of all shipments for rail and pipeline are of the same order of magnitude as those given above for refined products (2.341 kopecks for rail and .96 kopecks for pipeline) (Shul'ga, 1974, p.9), the figure for all shipments by water can be taken as roughly that for refined products.

A significant problem with this treatment of transportation costs is that it assumes that they vary linearly with distance, when in fact, they do not. Transport costs generally taper with distance, and Soviet railroad freight rates are no exception (Sagers and Green, 1985b). Thus the cost function utilized underestimates the "cost" of short hauls and overestimates that of long hauls. However, simple distance is preferred to a more complex non-linear cost function. This is because a physical measure of total costs for the system (ton-kilometers) is obtained, which can then be compared with the reported amount of freight turnover to determine how closely aligned actual transport effort is with our optimal level.

D. MODEL RESULTS

Analysis of the Soviet distribution system for refined products is undertaken in two parts. The first examines flows in the 1980 network, while the second examines the pattern of flows and refinery utilization with the addition of all planned refinery expansions of the 11th Five-Year Plan (1981-1985).

1. Modeled Flows

The modeled flows are shown in Figure 5.3. Because the consumption and production of petroleum products are relatively decentralized, the flows are relatively small. Most flows are less than three million tons. Only 46 arcs of the over 1600 arcs in the system have flows of over 10 million tons, and only 6 of these have flows of over 20 million tons. These few arcs with the larger flows are usually the initial outlet from a large refinery into the network. The largest single flow in the network is 30 million tons, from Syzran' to Penza (Figure 5.3). Not surprisingly, this is the initial outlet of the large complex of refineries in the Volga-Urals region to the major consuming centers in the western portion of the USSR.

Relatively small distribution areas for each refinery are delimited, as the flow pattern is predominately local in character. There are few long distance interregional flows (Figure 5.3).

However, there are some exceptions to the general pattern of local distribution. Products from the large Volga-Urals refineries (e.g., Kuybyshev, Ufa, Syzran') are shown as moving to the west and south considerable distances into central Russia, the Ukraine, and North Caucasus. Distribution from the Omsk refinery encompasses a large area in West Siberia, extending from the Urals to the Kuznetsk Basin, and from Surgut to southern Kazakhstan and Central Asia (to Dzhambul and Frunze)

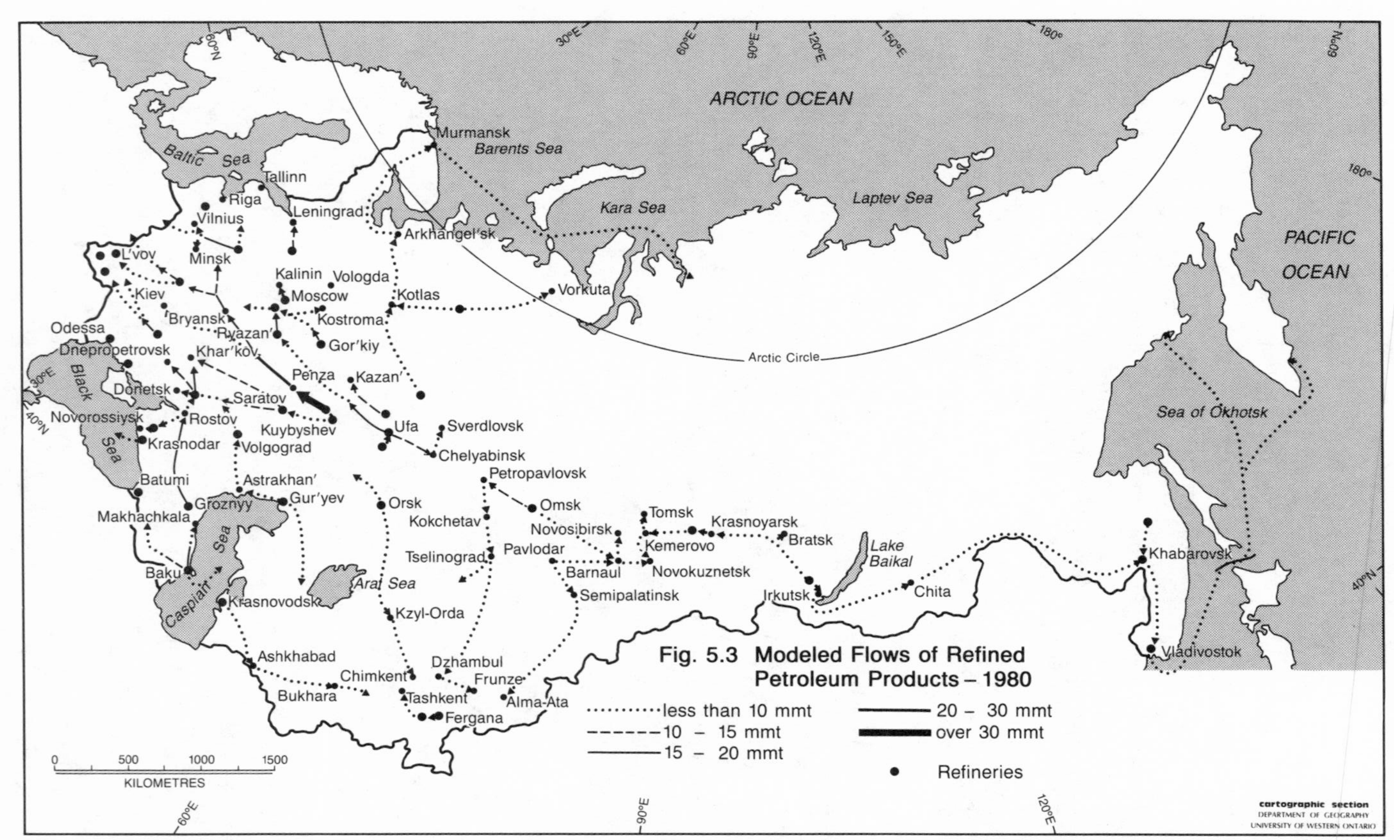

Fig. 5.3 Modeled Flows of Refined Petroleum Products – 1980

(Figure 5.3). The smaller refineries at Gur'yev and Krasnovodsk are the least-cost supply points for large areas on the eastern side of the Caspian Sea in Turkmenia, Uzbekistan, and Kazakhstan because of the sparse population and thus limited demand in the region. The Angarsk refinery also serves a huge area, extending from the cities of the Kuznetsk Basin to the Pacific (Figure 5.3). Interestingly, the Omsk and Angarsk refineries are mentioned as being far too large for the efficient distribution of refined products within their market areas (Oil and Gas Journal, 1984, p.26; Mozhin, 1983, p.5).

In general then, there is a rough congruence between the modeled pattern of flows and the actual pattern discussed above, despite the simplifying assumptions of the model. This is because mazut dominates the output mix, and its consumption is concentrated at electrical power stations and large industrial and municipal boilers in the major cities. These bulk flows evidently are captured fairly well by the gross aggregation of the model.

The localized flows of refined products in the optimal solution result from the assumption of a uniform output mix at all refineries and a product-undifferentiated consumption pattern. While an exact match between refinery output and the local consumption mix of products is not to be expected in reality because of the wide variety of specialized products produced by the refining industry, for overall efficiency in distribution, there still should be considerable agreement between the two. This is because it is so much cheaper to ship crude to the refinery than refined products from it.

An approximation of the magnitude of the discrepancy between refinery output and local consumption is the difference between the total ton-kilometers (transport costs) generated by the model and actual ton-kilometers incurred in moving refined products, as well as that between the actual and modeled average length of haul.

In 1980, the actual freight turnover of refined products in primary movements in 1980 was around 458 billion ton-kilometers. This includes about 364 billion ton-kilometers in railroad traffic, approximately 36 billion ton-kilometers on products pipelines, and other amounts of refined product freight traffic on river and sea tankers (about 40 and 16 billion ton-kilometers, respectively). However, model-generated costs amount to only about 50 percent of this total, or 228 billion ton-kilometers. Also, the average length of haul for refined products in 1980 for all modes of transport was 997 kilometers (Table 5.2), whereas that for the modeled flows is only 522 kilometers.

These significant differences between the modeled and actual flows result from several factors. One is that refined product consumption is more spatially dispersed than in the model; smaller cities, farms, and smaller TETs, not included in the model, must also be supplied with refined products, and this generates additional freight traffic not included in the model. Other modeling deficiencies are the use of straight-line distances between nodes and excluding non-refinery sources of petroleum products (e.g., gas refineries, topping plants). Also, a degree of specialization at a few refineries for certain products not utilized in large quantities in the Soviet economy (e.g., particular kinds of lubricants and additives) would be necessary for economies of scale. So products such as these would be shipped from a few production sites to consumers throughout the USSR.

Seasonal changes in consumption patterns constitute another factor for the discrepancy in total freight turnover. During the winter, more heating fuel and mazut are needed because of increased heating requirements and the switch by "buferniy" electrical stations from gas to mazut (Grigor'yev and Zorin, 1980, p.16). The general lack of reserve capacity in the refining industry has meant that the need for extra boiler fuel in winter has been met by reducing the amount of light products being converted from heavier fuels in catalytic crackers (Sagers and Tretyakova, 1985, p.22). This practice probably generates a considerable volume of extra traffic because catalytic cracking capacity is limited and at relatively few refineries (Sagers, 1984; Sagers and Tretyakova, 1985). In fact, much of this extra boiler fuel would have to be shipped all the way from the refineries at Omsk and Angarsk to the European USSR because the Siberian refineries are relatively well-equipped with secondary processing, while mazut accounts for a small share of the fuel balance of eastern electrical stations (Nekrasov and Troitskiy, 1981, p.232).

However, the discrepancy in total freight turnover between the model and reality is too large to be attributed entirely to the factors given above. Probably the major reason for the discrepancy is the mismatch between the structure of local consumption and refinery output. Apparently then, the gross aggregation of the model captures the main direction of bulk flows, particularly mazut, while the complex distribution pattern for the other products is not wholly represented.

An example of this apparently occurs in West Siberia. Although several sources indicate that products are shipped from the Volga-Urals refineries into West Siberia and northern Kazakhstan (Figure 5.2), the modeled flows do not exhibit this pattern (Figure 5.3). This discrepancy apparently is the result of the different consumption mixes in Siberia and the European USSR. In the European portion, electrical power stations and other boilers are the main consumers of petroleum products, so mazut is the main product needed. In contrast, gas and coal are more plentiful in Siberia to supply boiler needs, so more light products are needed to power transport and agricultural machinery (Sagers, 1984, p.8). Thus, mazut apparently is shipped from West Siberia while light products are shipped in. This type of cross-haul cannot be captured by the product-undifferentiated model.

The large discrepancy in overall freight turnover between the modeled and actual flows implies that a major problem in Soviet refining is the lack of proper planning to assure that production facilities for the various products needed in the local area are installed at the refinery in question. According to at least one Soviet source (Karchik, 1985, p.63), this is in fact a serious problem. Apparently then, production (supply) factors predominate in the decision concerning where to add new refining capacity, with insufficient attention being given to demand considerations. Such a bias is fairly typical throughout the Soviet economy (e.g., Davies, 1974).

The large discrepancy between modeled and actual flows in total turnover implies that the distribution of refined petroleum products is not very efficient, placing an unnecessary burden upon railroads. This is in addition to the USSR's existing inefficient geographical distribution of refining capacity, because the model takes as a given the location of the supply nodes in the solution.

The extra 228 billion ton-kilometers incurred in distributing refined products (the difference between the modeled and actual flows) represents an astounding 6.6 percent of the total freight turnover on Soviet railroads. In monetary terms, this extra volume of freight movement adds something on the order of 57 million rubles in operating costs to the economy, given an average shipping cost of about 2.5 kopecks per ten ton-kilometers on the railroads (Shafirkin, 1978, pp.222-223).

The difference between model-generated costs and those actually incurred imply that significant potential exists for reducing the transportation burden on the railroads by improving the match between refinery output and local consumption patterns. This would be in addition to the savings resulting from greater dispersion of refining capacity, an on-going development during the last two decades. Improving the match in output is possible with the planned transition of the industry to deeper refining (Sagers and Tretyakova, 1985). This program involves the expansion of secondary processing capacity, particularly cracking, to improve the product mix; i.e., producing more light products per ton of oil refined. More secondary processing capacity, particularly if located at the appropriate refineries, would provide greater flexibility in the output mix, and would allow it to be adjusted to better fit local consumption patterns and seasonal variations.

This inefficiency in distributing refined products is in contrast to the overall efficiency noted in a previous study of electricity (Sagers and Green, 1982) and in the previous chapters on gas and crude petroleum. Thus, the Soviet distribution networks operating near optimum efficiency are those which handle a fairly homogeneous commodity in a unique transportation system. Since refined products are distributed largely by a common carrier (the railroad) and are heterogeneous in nature, this extra complexity as well as the more fragmented system of decision-making for refined product shipments apparently must account for the inefficiency.

2. New Refining Capacity

Some indication of the most efficient location for new refining capacity can be inferred from the model because it provides an opportunity cost for each of the refineries as part of the optimal solution. These opportunity costs shown how much transport costs (for refined products) would be decreased if more capcity were available at each site.

These opportunity costs are given in Table 5.5. They indicate that the refinery with the highest opportunity cost is L'vov. At L'vov, an additional ton of capacity would decrease transport costs in the system by 25.9 million ton-kilometers. With an average cost of 2.5 kopecks per ten ton-kilometers for shipping refined products by rail, each additional ton of capacity at L'vov therefore would save about 6.5 million rubles annually in transport costs on refined products. This is probably more than enough to warrant construction since the cost of a new standard refining unit is 37 million rubles (Sredin and Lastovkin, 1972, p.33).

Differences in opportunity costs tend to be regionalized, reflecting the availability of supply relative to demand. The highest opportunity costs are at refineries in the western Ukraine (L'vov, Drogobych, and Nadvornaya) and the Far East (Khabarovsk and Komsomol'sk) (Table 5.5). The shortage of petroleum products in these regions is in fact mentioned by Danilov _et al._ (1983, pp.355, 383). Other areas with relatively high opportunity costs are the Baltic (Mazeikiai), eastern and

TABLE 5.5

Transport Opportunity Costs for Soviet Refineries in 1980
(million ton-kilometers per ton of extra capacity)

Rank	Refinery	Opp.Cost	Rank	Refinery	Opp.Cost
1	L'vov	-25.9	21	Ukhta	-9.4
2	Khabarovsk	-25.4	22	Krasnodar	-9.1
3	Drogobych	-25.3	23	Saratov	-9.1
4	Nadvornaya	-24.4	24	Tuapse	-7.6
5	Mazeikiai	-23.3	25	Syzran'	-6.7
6	Komsomol'sk	-22.8	26	Perm'	-6.3
7	Odessa	-21.3	27	Kuybyshev	-5.4
8	Mozyr'	-20.8	28	Nizhnekamsk	-5.3
9	Novopolotsk	-20.2	29	Groznyy	-5.0
10	Kirishi	-19.6	30	Pavlodar	-4.7
11	Kherson	-19.1	31	Orsk	-4.4
12	Kremenchug	-17.7	32	Batumi	-3.0
13	Lisichansk	-14.4	33	Gur'yev	-2.3
14	Moscow	-14.3	34	Salavat-Ishimbay	-1.7
15	Yaroslavl'	-14.1	35	Ufa	-1.7
16	Khamza	-13.5	36	Omsk	-1.6
17	Fergana	-12.8	37	Baku	0
18	Ryazan'	-12.5	38	Krasnovodsk	0
19	Volgograd	-10.4	39	Angarsk	0
20	Gor'kiy	-10.2			

southern Ukraine (Odessa, Kherson, Kremenchug, Lisichansk), and Belorussia (Mozyr', Novopolotsk). These areas with the highest opportunity costs would seem to be the most likely areas for refinery expansion in the 11th Five-Year Plan. They are mostly along the western border, and so partially reflect the demand for exports. In contrast, relatively low costs are found in the Volga, Caucasus, and Siberian regions.

Refinery expansion in the 11th Five-Year Plan is designed to reduce transportation costs by meeting the needs of deficit areas (Biryukov et al., 1982, p.30). The success of Soviet planning in achieving this goal can be evaluated by adding all new capacity planned for the 11th Five-Year Plan into the model. This includes additional primary refining units at Lisichansk, Mazeikiai, and Pavlodar, and the establishment of new refineries at Achinsk, Chimkent, and Neftezavodsk (Sagers, 1984). Since each unit has a capacity of about 7 million tons, this amount is added to the 1980 capacity of each (Table 5.3) in a subsequent model run.

In this second simulation, all the additional refinery capacity is fully utilized except for that at Neftezavodsk, where only 4.8 of the 7 million tons are drawn upon. Since the demand is at 1980 rather than 1985 levels, in general, the additional capacity at these particular sites seems warranted. Also, overall costs for the system drop by 23.7 percent to 184.1 billion ton-kilometers with the additions, a significant improvement in efficiency.

Although this new capacity alleviates some of the supply problems of the Ukraine, Baltic, Kazakhstan, and Central Asia, the most severe shortcomings previously identified in the system, in the western Ukraine and Far East, still remain. Therefore in the future, these are the areas most likely to receive new refinery capacity. In fact, expansion in throughput capacity may have occurred at Komsomol'sk in 1977 (Sagers, 1984, p.53). The need for new refinery construction in the Far East is reported by Danilov et al. (1983, p.355) and it was announced that work would start during 1981-1985 on a new refinery in Moldavia ("News Notes", SG-April 1981, p.275). Although nothing has appeared to suggest that work is progressing, Grigor'yev *et al.* (1984, p.60) imply that Moldavia and the Far East are being given first priority for refinery expansion.

However, it is important to note that about half of the crude petroleum needs of the refineries in the Far East are met with West Siberian crude railed in from Angarsk, rather than supplied by pipeline. Thus, the potential savings in transport costs for refined products resulting from the location of new refinery capacity in the region are not as great as they would be for other refineries. Thus, the new Achinsk refinery in East Siberia, which is on the main pipeline system, in fact may take the place of expanded refinery capacity in the Far East.

E. CONCLUSIONS

The surplus and deficit areas identified in the analysis generally correspond to those reported to exist. Also, the general pattern of model-generated flows corresponds roughly to the actual pattern after taking into account new refining capacity and changes in consumption. Thus, despite the gross aggregations and limiting assumptions in the model made necessary by the lack of data, the model provides a coarse approximation of refined product distribution in the USSR.

A second model simulation demonstrates that in general, the locations chosen for refinery expansion in the 11th Five-Year Plan appear justified. Although the pressing problem identified in the western Ukraine is not addressed, plans for the construction of a new refinery in Moldavia have been announced. Thus, the problems caused by the concentration of refining capacity in the traditional oil-producing regions are being rectified slowly as new refining capacity is constructed in the major consuming regions.

However, production and local consumption patterns for the various products do not coincide, resulting in unnecessary long hauls. The model and its optimal flows generate only about half the actual level of freight traffic, indicting that nearly half of all freight turnover results from cross-hauls between regions of various products as a result of this mismatch. Thus the analysis indicates that the distribution of refined products is relatively inefficient. This contrasts sharply with the efficiency noted in electricity, crude oil, and natural gas flows. This implies that Soviet planning is adequate for homogeneous commodities in unique transportation systems, but has problems dealing with the complexity of heterogeneous products on common carriers. However, this does mean that significant potential exists for relieving a considerable amount of the railroads' transportation burden by improving the match between refinery output mix and local consumption patterns.

VI
Soviet Coal Movements

A. INTRODUCTION

Coal long has been an important energy source in the USSR. Until the upsurge of the hydrocarbons in the 1950s, with petroleum supplanting coal as the USSR's principal energy source, coal played the principal role in Soviet energy supply, accounting for over 60 percent of all fuel produced (Narkhoz SSSR 1922-1972, p.162). However, by 1980, coal accounted for only 25.2 percent of Soviet production of primary energy, and in 1984, 22.8 percent (Narkhoz SSSR 1984, p.166). Although its relative contribution has declined because of the enormous increase in the production of petroleum and natural gas, the production of coal has increased in absolute terms (Table 6.1), from 509.6 million tons of raw coal in 1960 to 716.4 million tons in 1980, and 712.3 million in 1984, or in calorific equivalents, from 373.1 million tons of standard fuel in 1960 to 484.4 million tons in 1980, and 476.8 million in 1984 (Narkhoz SSSR 1984, p.166).

As the deteriorating situation for petroleum became evident in the mid-1970s, initially energy policy favored accelerated development of coal because of the country's huge coal reserves (Gustafson, 1982; Dienes and Shabad, 1979, p.103). However, severe problems in the coal industry led to an unprecedented decline in production in the late 1970s, stemming largely from the lag in development of new mine capacity to replace mine depletions ("News Notes", SG-April 1982, pp.287-290). This forced a change in strategy to gas (Gustafson, 1982). The eclipse of the coal strategy was due to a combination of factors, the most important probably being the history of underinvestment in the industry and its resulting poor technological modernization (Office of Technology Assessment,

TABLE 6.1

Soviet Coal Production
(million tons)

	1970	1975	1980	1985
USSR	624	701	716	726
European Russia	89.5	94	85	116
Moscow Basin	36	34	26	19.3
Pechora Basin	21.5	24	28	29.8
Donets Basin (east)*	32	33	31	31.0
Urals	53.5	45	41	36.0
Siberia	199	242	265	276
Kuznetsk Basin	113	139	145	141.6
Kansk-Achinsk Basin	18	28	35	40.8
So. Yakutian Basin	—	—	2.5	10.0
Other	68	75	82.5	83.6
Ukraine	207	216	197	190
Donets Basin*	216	222	204	197.1
L'vov-Volyn Basin	12	14.5	n.d.	15.0
Dnieper Basin	11	13	n.d.	15.0
Kazakhstan	62	92	115	130.8
Karaganda	38	46	49	49.8
Ekibastuz	23	46	67	80.5
Central Asia	8	10.5	11	11.0
Uzbekistan	4	5	6	6.0
Kirgizia	4	4	4	4.0
Tadzhikistan	neg.	1	1	1.0
Georgia	2	2	n.d.	2.0

Sources: "News Notes", SG-April, 1985, p.301; Dienes and Shabad, (1979, pp.110-111); Ugol', no. 3, (1986), pp. 57-61.

*Also portion of Donets Basin in RSFSR (east wing in Rostov Oblast).

1981; CIA, 1980; Gustafson, 1982). These conditions made it next to impossible for a substantial increase in output to be achieved in less than a decade because of the long lead times required to bring new mine capacity on line (CIA, 1980). Other considerations included the relatively high cost of the new mine capacity as well as the fairly limited potential for substituting coal for oil (CIA, 1980).

There is also a geographical element in the demise of the coal strategy. The accessible coal reserves of the European USSR (principally in the Ukraine's Donets Basin) have been seriously depleted, while most of the USSR's huge reserves are located in the east (Figure 6.1). Therefore coal production has been shifting eastward, although the eastern reserves are generally of lower quality, expensive and difficult to develop, and require considerable transport to get the energy to the main sources of demand in the European USSR. Whereas the European USSR produced 56 percent of the USSR's coal in 1970, and 45 percent in 1980, the proportion was 47 percent in 1985 (Table 6.1).

As a result, east-west shipments of coal have been steadily rising, from 66 million tons in 1970 to 96 million in 1975. By 1980, they were around 120 million tons ("News Notes", _SG_-December 1982, pp.767-770). Of this, the amount shipped to the regions west of the Urals was planned to reach about 46 million tons in 1980 (Mitrofanov, 1977, p.83).

It is principally the railroads which have the burden of transporting coal in the USSR. Coal loadings on railroads in 1980 were 731.6 million tons, while river and sea shipments were only 23.8 and 9.6 million tons, respectively (_Narkhoz SSSR 1980_, pp.296, 299, 302). Coal is the single most important commodity in generating rail freight traffic, in 1980 accounting for 17.4 percent of the Soviet railroads' total ton-kilometers and 20 percent of the tons originated (_Narkhoz SSSR 1980_, p.195).

Therefore the growing volume of long-distance coal hauls from the eastern basins, necessitated by stagnating or declining production in the Donets Basin and other western coal-producing areas, is

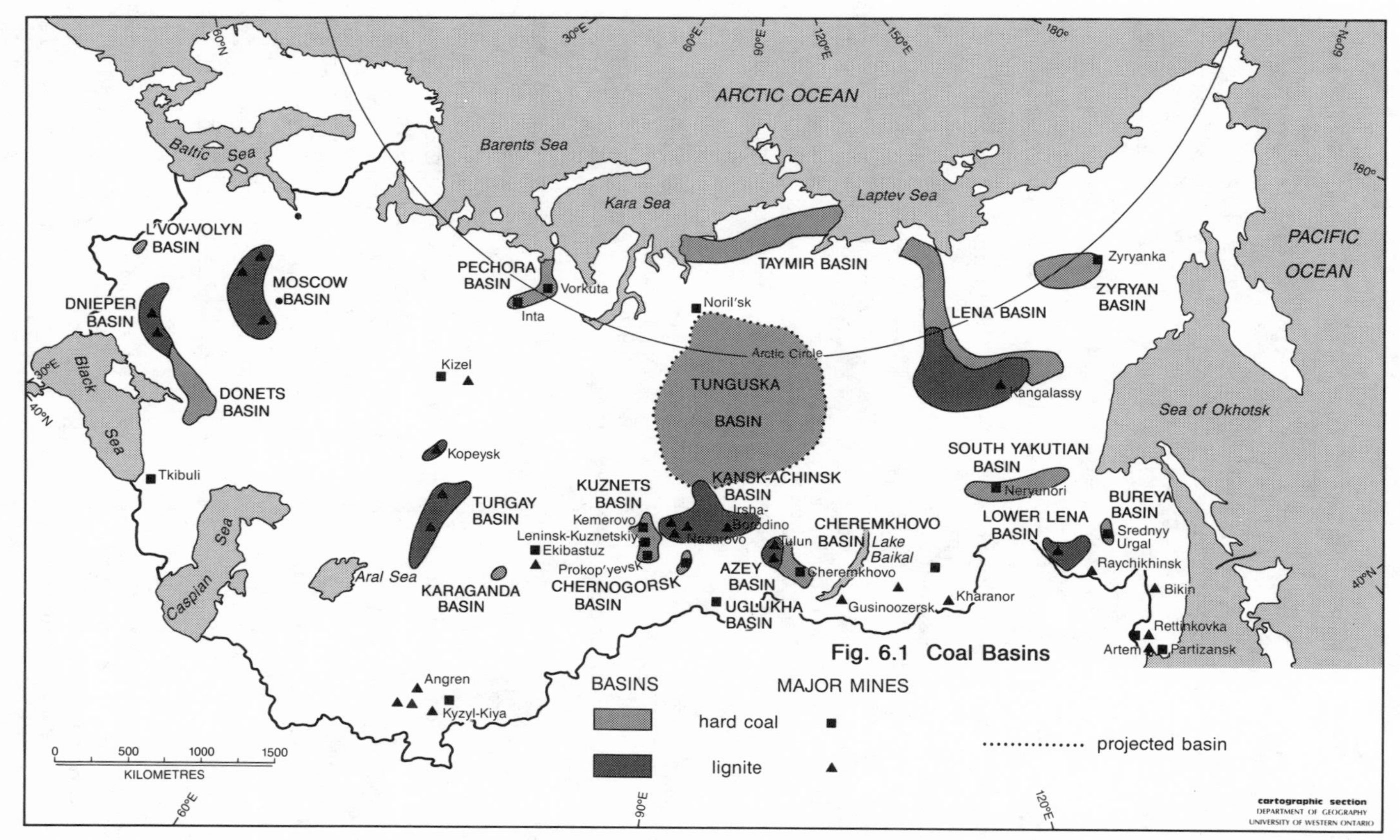

Fig. 6.1 Coal Basins

placing an increasing burden on an already overloaded rail system (Hunter and Kaple, 1983). Reflecting this fact, the average length of coal hauls by rail increased from 692 kilometers in 1970 to 695 kilometers in 1975, and then jumped to 818 kilometers in 1980 ("News Notes", SG-December 1982, p.771); in 1984 the average length of haul was down slightly, to 808 kilometers (Narkhoz SSSR 1984, p.338).

This chapter examines the problem of transporting coal in the USSR. Since the production and distribution of coking coal for 1980 are analyzed in considerable depth by ZumBrunnen and Osleeb (1986), the focus here is upon movements of non-coking coal. This is mostly steam ("energeticheskiy") coal for boiler-furnace use, although some is used for technological purposes. Most of steam coal is used for generating electrical power, although 239 million tons of raw coal were apparently utilized by other industrial users (including small electrical stations not operated by the Ministry of Electrical Power) and the household-communal sector in 1980. This amount for non-electrical station consumption of steam coal is calculated as gross mine output (716 million tons: Narkhoz SSSR 1980, p.157), less coking coal production (178 million tons: CIA, 1982, p.205), less deliveries to central electrical power stations (299 million tons: Nekrasov and Troitskiy, 1981, p.225).

Included within the 239 million tons of apparent consumption are the considerable losses involved in beneficiating, other processing, transporting, and storing coal. Because available production data by basin are in gross output and losses occur mostly for coking coal, initially losses are not considered in our transportation model.

However, the difference between gross mine output and marketable (net) output alone is significant. This mostly affects hard coal, as these losses largely result from beneficiation. Gross mine output of hard coal in 1980 was 553 million tons ("News Notes", SG-April 1982, p.288), while net (marketable) output was only 493 million tons (SEV, 1981, p.81), a difference of 60 million tons. In contrast, the difference for lignite is much less because it generally is not beneficiated. Gross mine output of lignite in 1980 was 163 million tons, with a net output of 160 million tons ("News Notes", SG-April 1984, p.277; SEV, 1981, p.81). Since almost all of coking coal is beneficiated, but only 20 percent of non-coking coal is (Tretyakova, 1985), most of the losses for hard coal are actually restricted to coking coal. For example in 1975, 68 percent of the difference between gross and net output of hard coal as a whole was accounted for by coking coal losses (SEV, 1976, p.77; Dienes and Shabad, 1979, p.104).

Another major source of losses is transportation. Open railroad cars often lose a ton per car for every thousand kilometers travelled (Office of Technology Assessment, 1981, pp.97-98). In 1975, transportation losses for Coal Ministry shipments were 35.6 million tons, about 5 percent of its production (P.K. Dubinskiy, 1977, p.27).

B. ABSTRACTING THE COAL DISTRIBUTION SYSTEM INTO A NETWORK

Although only limited information is available on regional coal production and consumption levels and transport capacity, the Soviet coal distribution system of 1980 is modeled. The network consists of supply nodes (coal fields), demand nodes (coal-fired power stations and large cities), and bounded arcs (railroad lines and some shipping routes). Evaluation of the general pattern of non-coking coal movements and mine and railroad utilization is undertaken for the distribution network.

1. Supply Nodes

Supply nodes in the network are the Soviet coal fields or coal-producing areas (Figure 6.1). A total of 31 supply nodes are utilized. Five are used to represent the main areas of the Donets Basin (Donetsk, Novoshakhtinsk, Voroshilovgrad, Gukovo and Pavlograd), with two nodes each for the Kuznets Basin (Mezhdurechensk and Kiselevsk), Kansk-Achinsk Basin (Irsha-Borodino and Nazarovo) and Pechora Basin (Vorkuta and Inta) (Table 6.2). The supply availability at each is determined by the basin's

TABLE 6.2

Coal Supply Nodes in 1980
(without coking coal)

Basin	Town	Amount (million tons)
Donets	Donetsk	62
Donets	Novoshakhtinsk	20
Donets	Voroshilovgrad	36
Donets	Gukovo	1
Donets	Pavlograd	11
Moscow	Novomoskovsk	25
Pechora	Vorkuta	1
Pechora	Inta	9
Urals	Kizel	3
Urals	Kopeysk	20
Urals	Volchansk	7
Kuznetsk	Mezhdurechensk	24
Kuznetsk	Kiselevsk	65
Kansk-Achinsk	Irsha-Borodino	19
Kansk-Achinsk	Nazarovo	15
So. Yakutia	Neryungri	3
Karaganda	Karaganda	22
Ekibastuz	Ekibastuz	67
Uzbek	Angren	6
Kirgiz	Kyzyl-Kiya	4
Georgia	Tkvarcheli	0
L'vov-Volyn	Chervonograd	15
Dnieper	Aleksandriya	10
Cheremkhovo	Cheremkhovo	12
Azey	Azey	12
Gusinoozersk	Gusinoozersk	2
Raychikhinsk	Raychikhinsk	15
Luchgorsk	Luchegorsk	13
Artem	Artem	5
Chita	Chita	8
Ussuriysk	Ussuriysk	6
Other	(Mezhdurechensk)	20
TOTAL		538

Sources: "News Notes," April 1982, pp.287-290; May 1979, p.332; February 1979, p.126; September 1982, pp.540-543, 545-548; March 1983, p.252; Dienes and Shabad (1979, pp.103-130), CIA (1982, p.205).

annual output of raw coal less coking coal production.

Coking coal production for 1980 in each basin is taken from the estimates given by the CIA (1982, p.205). Coking coal production in the Donets Basin (74 million tons) is assumed to occur mostly at the Donetsk and Voroshilovgrad nodes (32 million tons each), with the remainder at Gukovo. All coking coal in the Pechora Basin (18 million tons) is assumed to be produced at Vorkuta, and in the Kuznetsk Basin, most of the coking coal is assumed to originate at the Mezhdurechensk node (48 out of 55 million tons). Twenty million tons of coal are produced at a number of unknown smaller fields. This amount is allocated to a dummy node at Mezhdurechensk, as it is located near the center of the network.

Removing coking coal from the analysis eliminates one of the most serious problems for the model, that of qualitative differences in the coal. The problem still exists, as the coal of the various basins remains heterogeneous, but coking coal is the highest quality coal and in extremely short supply in the USSR (CIA, 1982; Izvestiya, July 6, 1984). Its use in the metallurgical industries, particularly ferrous metallurgy, also sets it apart. In contrast, steam or "energeticheskiy" coal is used basically as a boiler fuel for generating heat and electricity.

The remaining qualitative differences among the various coals should not pose too great a problem for the model. The low-quality coals (e.g., Moscow and Kansk-Achinsk lignites) are not very transportable and so mostly are used at nearby electrical power stations. Since the model allocates available supply to the nearest demand nodes, it should match the lignite with the nearby stations.

Two exceptions to this are the Ekibastuz and Angren coals. The subbituminous coal from Ekibastuz in northern Kazakhstan is being hauled on an increasing scale to electrical stations in the Urals, a distance of over 1400 kilometers. In 1980, Urals electrical stations used 33 million tons of Ekibastuz coal, and since 1978, the Cherepet' station in Tula Oblast near Moscow also has been using Ekibastuz coal (Nekrasov and Troitskiy, 1981, p.228), a distance of about 3000 kilometers. The Cherepet' station had been using local lignite from the Moscow Basin, but reserve exhaustion in the Moscow Basin forced it to begin using Karaganda coal in 1976 and then Ekibastuz coal in 1978 ("News Notes", SG-December 1982, pp.771-772).

Declining output in the Moscow Basin also seems to be the reason for the shipment of Angren lignite from Central Asia to help fuel the 1200 MW of coal-fired capacity at the Ryazan' station ("News Notes", SG-December 1982, p.772; March 1982, p.197), a distance of about 3300 kilometers. The Angren shipments are expected to cease once a new station under construction in the Angren Basin itself goes into operation in 1984 (Biryukov *et al.*, 1982, p.33).

Anthracite coal is also qualitatively different from other steam coals. Almost all of the USSR's anthracite (about 75 million tons annually) is produced in the Donets Basin, in Rostov Oblast around Novoshakhtinsk and in neighboring Voroshilovgrad Oblast in the Ukraine. Much of the anthracite is used to fuel local electrical stations, although a considerable amount is used as a boiler fuel at a number of industrial enterprises throughout the country (e.g., Donets anthracite is used as far away as the Kuznetsk Basin; see Biryukov *et al.*, 1982, p.22), as well as for heating in the household-communal sector (Tretyakova, 1980).

2. Demand Nodes

Demand nodes in the system consist of major coal-fired electrical stations and the USSR's large cities. A total of 55 electrical stations are included as demand nodes in the study. Total coal consumption by central electrical power stations in 1980 was 299.3 million tons, supplied from several different basins (Table 6.3). Coal consumption for each of the 55 stations generally is determined by allocating the amount for each basin or region among the appropriate stations according to installed capacity. However, only half of the capacity of stations utilizing coal part of the year (dual-fired

TABLE 6.3

Coal Consumption by Central Electrical Power Stations
(Deliveries)

Basin	Amount (million tons)
TOTAL	299.3
Ekibastuz	62.4
Kansk-Achinsk	23.3
Kuznetsk (open-pit)	39.7
Donets	59.7
Moscow	21.8
Chelyabinsk (Kopeysk)	10.1
Kizel	2.1
L'vov-Volyn	7.3
Other	72.9
of this:*	
Angren	5.7
East Siberia/Far East	19.0
Karaganda	18.5
Pechora (Inta)	1.0
Kuznetsk (deep-mined)	35.0

*Disaggregation of the "other" deliveries are only estimates of seems probable given the installed capacity of the stations being served by the various coal fields, coking coal deliveries, and gross output levels.

Sources: Nekrasov and Troitskiy (1981, pp.225,228); "News Notes," SG-December 1982, p.770; April 1982, pp.287-290; September 1982, pp.540-543, 545-548.

or "bufferniy" stations) is used in determining the allocation. This allocation of coal among the stations results in a regional distribution conforming closely to estimates based upon the percentage data given in Nekrasov and Troitskiy (1981, p.232), once converted to standard fuel equivalents.

Deliveries of Ekibastuz coal to central electrical stations in 1980 were 62.4 million tons (Table 6.3). Of this, 53.4 percent (about 33 million tons) went to stations in the Urals (Nekrasov and Troitskiy, 1981, p.228), and perhaps 5 million tons went to nearby West Siberian stations, with maybe a million tons being shipped to the Cherepet' station in Tula Oblast (see below). The remainder (23 million tons) represents consumption within Kazakhstan. In 1979, 32 million tons of Ekibastuz coal went to the Urals and 3 million tons to West Siberia, with 22 million tons remaining in Kazakhstan ("News Notes", SG-March 1980, p.189). Shipments to Cherepet' in 1979 are not mentioned, although they began in 1978, so usage must be fairly small.

The stations in the Urals using Ekibastuz coal are known to include Reftinsk, Troitsk, and the Kurgan TETs (heat and power station), as well as Verkhne-Tagil and Serov (Nekrasov and Troitskiy, 1981, p.228; P.K. Dubinskiy, 1977, p.36), so 33 million tons are allocated among them. The West Siberian stations receiving Ekibastuz coal evidently include TETs at Omsk and Barnaul, and 5 million tons are allocated between them. The stations in Kazakhstan fueled by Ekibastuz coal are Ekibastuz, Yermak, Karaganda, and TETs at Pavlodar and Petropavlovsk ("News Notes", SG-March 1980, p.188). Only half of Karaganda's capacity is considered in the allocation since it also uses Karaganda coal. The allocation of 23 million tons among them results in the 1000 MW Ekibastuz station (its capacity in 1980) receiving a little over 4 million tons, more or less the 4 million tons per 1000 MW as planned (Dienes and Shabad, 1979, p.116).

In 1980, 23.3 million tons of Kansk-Achinsk coal were delivered to central electrical power stations (Table 6.3). This would be to the station at Nazarovo and the Krasnoyarsk complex of thermal stations, and is allocated accordingly. Apparently, a small amount also is used by West Siberian stations, such as the Novosibirsk TETs (P.K. Dubinskiy, 1977, p.23).

Deliveries of Kuznetsk coal to electrical power stations in 1980 included 39.7 million tons mined in open pits, plus about 35 million tons of deep-mined coal (Table 6.3). Twelve million tons of Kuznetsk coal were shipped to the European USSR (Nekrasov and Troitskiy, 1981, p.228), leaving about 63 million tons for use in the Urals and West Siberia. This amount is allocated among the Verkhne-Tagil, Yuzhnoural'sk, Tom-Usa, Belova, and South Kuzbas stations, and the Novosibirsk and Kemerovo TETs. Since the Yuzhnoural'sk station probably also uses local Urals coal and the Verkhne-Tagil station also uses Ekibastuz coal, gas, and oil, only half of their capacity is utilized in determining the allocation (see below). This allocation provides 16 million tons of Kuznetsk coal to Urals electrical stations.

At the Karaganda Basin, total production was about 49 million tons in 1980, of which 27 million tons was coking coal (CIA, 1982, p.205). Apparently 18.5 million tons were delivered to central electrical stations (Table 6.3), of which perhaps 6 million tons went to the European USSR (see below), leaving 12.5 million tons for consumption by the stations at Karaganda itself.

Lignite deliveries from the Moscow Basin in 1980 were 21.8 million tons (Table 6.3). The major stations which have traditionally used Moscow coal are Kashira, Ryazan', Shchekino, Novomoskovsk, and Cherepet'. But in the 1970s, reserve exhaustion in the Moscow Basin led to declining production. As a result, Kuznets coal was supplying much of the needs of the Kashira and Cherepet' stations, as well as the Moscow TETs in 1975 (P.K. Dubinskiy, 1977, p.33). Cherepet' was forced to switch to Karaganda and Ekibastuz coal after 1976, and evidently about 3 million tons of Angren lignite were used at Ryazan' in 1980, supplying about half of its needs. The amount of coal utilized from the Moscow Basin in 1970, 1975, and 1980, when compared with the aggregate capacity of the main stations it supplied, indicates that 4-5 million tons of lignite are needed for each 1000 MW of capacity. Thus there is a shortage of about 3 million tons in 1980, which is assumed to represent the shipments of Angren coal to Ryazan'. This shipment still leaves 2-3 million tons of coal production at Angren still available to fuel the existing 600 MW Angren station in Central Asia. Consumption of Angren coal is planned to be 4.6 million tons in 1985 with the startup of the new Angren station of 1200 MW now under construction (Nekrasov and Troitskiy, 1981, p.235; Biryukov et al., 1982, p.33). Therefore the 21.8 million tons delivered from the Moscow basin are allocated among these stations utilizing only half of Ryazan's coal-fired capacity and excluding Cherepet'.

Coal deliveries from the L'vov-Volyn Basin to power stations were 7.3 million tons in 1980 (Table 6.3). Only two stations, Dobrotvor and Burshtyn, are supplied from this source (Danilov et al., 1983, p.383). Since the 2400 MW Burshtyn station also burns oil, only half of this capacity is used in determining the allocation.

Deliveries of coal from the Urals' Kizel and Chelyabinsk Basins in 1980 were 12.2 million tons (Table 6.3). This is allocated among the stations at Yuzhnoural'sk, Serov (utilizing only half their installed capacity since they were also considered in the allocation of Kuznetsk coal), Yayva, and TETs at Chelyabinsk, Sverdlovsk, and Sterlitamak.

Deliveries to central power stations from the coal fields in the Far East and East Siberia evidently were around 19 million tons in 1980 (Table 6.3), out of a total gross production of over 43 million tons ("News Notes", SG-April 1982, p.288). Thus 19 million tons are allocated among the Raychikhinsk, Primorsk, Artem, Vladivostok, Partizansk, Irkutsk, Gusinoozersk, Chita, and Ulan-Ude stations.

Coal consumption for the electrical stations in the European USSR is somewhat difficult to determine because coal is used from such a variety of basins. Their total cost consumption in 1980 was equivalent to 64.2 million tons of standard fuel, 33.9 million tons of which was in the Ukraine (Nekrasov and Troitskiy, 1981, p.228).

Donets coal deliveries were 59.7 million tons (Table 6.3), equal to approximately 37.1 million tons of standard fuel with the degradation of quality in recent years (Nekrasov and Troitskiy, 1981, p.224). Twelve million tons of Kuznetsk coal were utilized in the European USSR west of the Urals (about 9.1 million tons of standard fuel), as well as 7.3 million tons of L'vov-Volyn coal (about 4.5 million tons of standard fuel) and 21.8 million tons of Moscow coal (about 6.2 million tons of standard fuel). Apparently, with deliveries of about 3 million tons of Angren coal, 1 million tons of Ekibastuz coal, and 6 million tons of Karaganda coal, only about 1 million tons of Pechora coal were used by European central electrical stations. These amounts convert to about 7.2 million tons of standard fuel, which with the deliveries of Donets, Kuznetsk, L'vov-Volyn, and Moscow coal, sum to the 64.2 million tons of standard fuel consumed.

Declining deliveries of Donets coal and its lower quality have led to increasing shipments of Kuznetsk, Karanganda and Inta (Pechora) coals to stations in the Ukraine (Nekrasov and Troitskiy, 1981, p.228). At the same time, shipments to Donets coal northward continue, to supply TETs in the Moscow-Yaroslavl' area, counter to the flow of Kuznetsk coal (Nekrasov and Troitskiy, 1981, p.229).

The allocation of Moscow, Ekibastuz, L'vov-Volyn, and Angren coal among the European stations was done previously (see above). Therefore the deliveries of these coals (33 million tons) was subtracted from the European total, leaving 78 million tons. This amount is allocated among the remaining coal-fired stations in the European USSR according to installed capacity. The stations included are Moldavia, Pridneprovsk, Cherepet' (for Karaganda coal), Kurakhova, Gottwald (Zmiyev), Starobeshchevo, Novocherkassk, Tripol'ye, Voroshilovgrad, Ladyzhin, Uglegorsk, Zaporozh'ye, Krivoy Rog, Mironovskiy, Slavyansk, and TETs at Moscow and Kiev.

The remainder of <u>apparent</u> domestic coal consumption, 239 million tons, is used to fuel boilers and furnaces at a variety of industrial enterprises and the household-communal sector.

Although it also includes losses, this amount is allocated among large Soviet cities according to population size. City size is used as it is known to reflect the relative industrial importance of the city as well as the size of its housing and communal economy (Huzinec, 1978), the two largest consumers of coal after electric power (Kurtzweg and Tretyakova, 1982).

A total of 147 cities are included, comprising almost all oblast centers and those cities with over 100,000 inhabitants. An exception to this is the cities of the Caucasus, Baltic, and Central Asia. Since coal plays a relatively minor part in supplying the industrial, municipal, and household heating needs of these regions (see Gankin et al., 1972, p.78), only the republic capitals are considered in making the allocation. The city allocations are aggregated by region in Table 6.4.

No data are available to determine the actual overall total or the actual regional distribution of this consumption. However, in 1975, the Coal Ministry delivered 81.7 percent of its gross production to consumers (567.3 out of 694.6 million tons) (P.K. Dubinskiy, 1977, p.14). Of this, 271.5 million tons were delivered to central electrical stations and 125.4 million tons to enterprises of the ferrous metallurgy industry (P.K. Dubinskiy, 1977, p.35). The residual, 170.4 million tons, would be deliveries to other industrial users and the household-communal sector. If a similar proportion for deliveries is assumed from 1980's gross output of coal, and assuming that all coking coal is used in the metallurgical industries, only 108 million tons remains for deliveries to the other users after subtracting deliveries to electrical power stations. This is less than half of apparent deliveries (239 million tons) derived as a residual. This would imply an astounding figure for losses, so it seems unlikely. Data on all shipments from the Coal Ministry to the various regions in 1970 and 1975 are given in Table 6.5 for comparison.

In addition to these supply and demand nodes, over 100 intermediate nodes are added to the network to allow for junctions and proper geographic alignment in the transportation system. All the intermediate nodes are cities or towns.

TABLE 6.4

Estimated Coal Deliveries in 1980
(without coking coal -- million tons)

Region	All Deliveries	To Central Power Stations	To Other Users
Northwest	21.3	0	21.3
Center	75.4	34.3	41.1
Volga-Vyatka	8.1	0	8.1
Cen. Chernozem	6.0	0	6.0
Volga	20.6	0	20.6
Urals	78.4	61.2	17.2
No. Caucasus	12.9	3.8	9.1
West Siberia	65.6	50.8	14.8
East Siberia	35.5	28.3	7.2
Far East	11.5	6.0	5.5
Ukraine	117.2	74.2	43.0
Belorussia	8.4	0	8.4
Baltic	6.2	0	8.4
Transcaucasus	9.1	0	9.1
Central Asia	11.1	2.7	8.4
Kazakhstan	43.8	32.4	11.4
Moldavia	7.9	6.5	1.4

Sources: Aggregated from city and electrical station allocations.

TABLE 6.5

All Coal Deliveries
(million tons)

Region	All Deliveries		Donets		Kuznetsk		Karaganda		Ekibastuz	
	1970	1975	1970	1975	1970	1975	1970	1975	1970	1975
Northwest	29.5	28.4	3.8	1.0	6.5	7.8	0	0	0	0
Leningrad c.	2.9	1.7	.9	.2	1.6	1.2	0	0	0	0
Leningrad O.	3.9	4.2	1.6	.4	1.7	3.0	0	0	0	0
Central	47.0	51.3	10.7	5.0	4.3	11.4	neg.	neg.	0	0
Moscow c.	2.8	3.6	1.9	1.4	.9	2.3	0	0	0	0
Moscow O.	9.7	8.7	3.6	1.2	1.5	3.2	0	neg.	0	0
Tula O.	21.3	21.7	1.8	.2	.3	2.3	0	0	0	0
Cen. Chernozem	13.8	14.6	8.5	7.5	1.9	3.4	neg.	.4	0	0
Volga	17.4	14.9	4.1	2.9	5.9	3.9	3.3	3.6	0	0
No. Caucasus	13.2	14.9	13.0	14.8	neg.	neg.	0	0	0	0
Urals	81.8	80.1	.2	.2	21.7	20.9	9.8	9.5	10.4	23.6
West Siberia	60.7	75.4	neg.	neg.	51.9	62.8	0	neg.	1.5	2.9
Ukraine	144.5	165.5	127.0	139.1	neg.	1.7	0	2.1	0	0
Central Asia	10.2	13.0	neg.	neg.	.4	.6	1.9	3.0	0	0
Kazakhstan	42.3	55.0	neg.	neg.	9.2	10.1	20.9	24.4	11.0	19.3
Belorussia	6.5	3.1	5.0	1.8	0	0	0	0	0	0
Moldavia	4.5	5.6	4.1	5.2	0	0	0	0	0	0
Transcaucasus	3.6	3.2	1.8	1.6	0	0	0	0	0	0
East Siberia	46.4	57.9	0	neg.	.4	.2	0	0	0	0
Far East	30.6	36.0	0	0	0	0	0	0	0	0
TOTAL	591.5	657.8	197.9	197.8	108.5	129.1	36.1	43.4	23.0	45.8

Source: P.K. Dubinskiy (1977, p.32).

3. Bounded Arcs

Bounded arcs represent the rail lines and shipping routes in the system. The alignment of the rail lines is taken from several Soviet maps and atlases showing the railroad network (e.g., Atlas zheleznikh dorog SSSR, 1984). Arcs representing shipping routes are added to the system for cities not served by rail, such as between Magadan and Vladivostok and Petropavlovsk-Kamchatskiy and Vladivostok. No data are available to determine annual transmission capacities for the arcs so the model is allowed to determine the optimal flows practically unhindered by upper capacity constraints. A limit of 70 million tons is set as the upper bound initially (i.e., maximum transmission capacity) for all arcs, as this exceeds the output of any single supply node. Only a few arcs, such as Ekibastuz-Tselinograd, are required to transmit more than this in order to achieve a feasible solution.

Distance is used as a surrogate for transportation cost in the model. There is a close correlation between such costs and distance in the USSR, although the actual railroad freight rate structure is more complicated than such a simple linear relationship (Sagers and Green, 1985b).

An exception to this is the arc connecting the Ekibastuz coal field with the Ekibastuz electrical station. This arc is made 2000 kilometers long to simulate the lower transportability of the subbituminous coal at Ekibastuz in comparison with the higher quality coal of the Kuznetsk Basin. The lower quality of Ekibastuz coal largely restricts its use to only as far west as the Urals, whereas Kuznetsk coal has a wide market area in the European USSR. But because the Ekibastuz Basin is located at an intermediate point between the European USSR and the Kuznetsk Basin, the model is likely to utilize coal from the closer basin at Ekibastuz for European consumption and supply the Urals with Kuznetsk coal. The 2000 kilometer penalty should prevent this from occurring.

C. MODEL RESULTS

The modeled coal flows are shown in Figure 6.2. The pattern is dominated by the large flows from the eastern basins to the west. The largest flows in the system are in Kazakhstan, between Ekibastuz and Tselinograd and from Tselinograd to Kokchetav (90 million tons each). This is due to the large flow to the Urals from the Ekibastuz Basin being joined by a smaller flow of Kuznetsk coal coming through Barnaul. Then at Tselinograd, this flow is joined by coal moving north from Karaganda. These large flows of 90 million tons annually translate into about 4500 carloads per day (see Dienes and Shabad, 1979, p.117), obviously straining railroad facilities to the limit. In actuality, only about 3000 carloads per day leave Ekibastuz ("News Notes", SG-April 1984, p.282), and only 80 percent of this can be accommodated by the South Siberian line through Tselinograd (Kontorovich, 1982, p.55). This would translate into 2600-2700 carloads of Ekibastuz coal per day through Tselinograd, as the route also must accommodate some shipments of coal from the Kuzbas (Kontorovich, 1982, p.vi). Thus, the model shows that optimally, more coal should be shipped through there, and it demonstrates that considerable congestion exists on the South Siberian line through Ekibastuz.

Another major flow of 70 million tons is generated from the Kuzbas along the Trans-Siberian and Central Siberian lines to the Urals (Figure 6.2). This massive flow of coal westward must also be straining existing railroad facilities, especially considering the need to also move coking coal from the Kuzbas (Table 6.6). Interestingly, the Kuzbas flow initially follows the Central Siberian route through Barnaul to Omsk, rather than the Trans-Siberian through Novosibirsk. while this is not entirely correct, as currently most of the flow is along the Trans-Siberian line, it does support the proposals for turning the Central Siberian into a specialized line for coal shipments (Kontorovich, 1982, pp.xv, 114-115; "News Notes", SG-April 1984, p.280).

The model generates flows of eastern coal through the Urals to the European USSR, into the

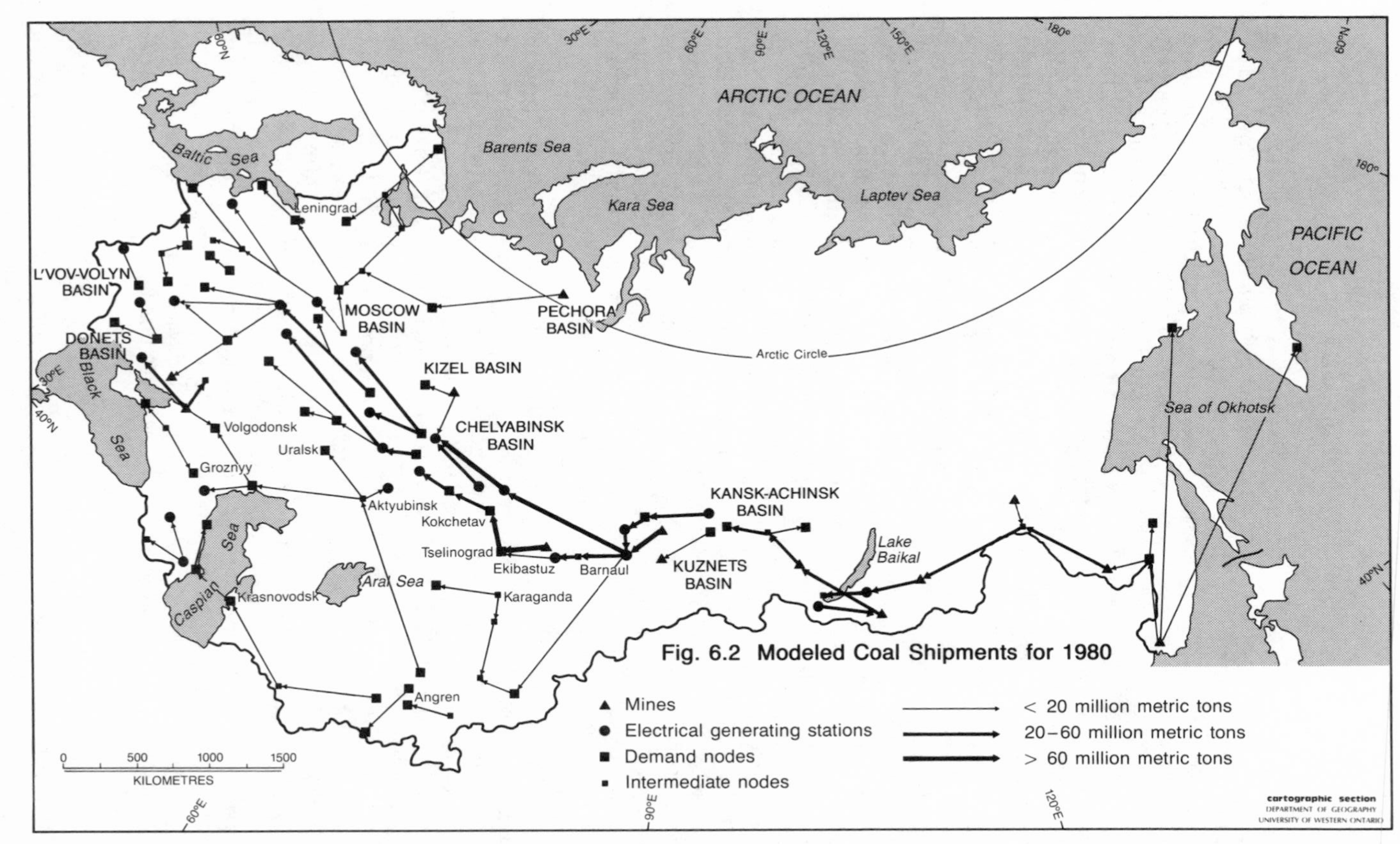

Fig. 6.2 Modeled Coal Shipments for 1980

TABLE 6.6

Approximate Structure of East-West coal Flows in 1980
(amount in million tons)

Origin	Destination	Amount	Purpose
Ekibastuz	Urals	33	steam
Ekibastuz	European USSR	1	steam
Angren	European USSR	3	steam
Kuznetsk	European USSR	12	steam
Kuznetsk	European USSR	18	coking
Kuznetsk	Urals	16	steam
Kuznetsk	Urals	17	coking
Karaganda	Urals	14	coking
Karaganda	European USSR	6	steam
	TOTAL	120	

Note: ZumBrunnen and Osleeb (1986) suggest a flow of 3 million tons of coking coal from Karaganda to the Urals, a flow of 9 million tons of coking coal from the Kuzbas to the Urals, and a flow of 9 million tons of coking coal to the European USSR from the Kuzbas for use in ferrous metallurgy.

Sources: Nekrasov and Troitskiy (1981, pp.225, 228); "News Notes," December 1982, p.770; May 1982, p.374; September 1982, p.546; estimates in text.

Ukraine, central Russia, the Baltic, and Belorussia. Almost all of the European USSR is shown as being supplied with eastern coal, the exceptions being areas in the Ukraine in the immediate vicinity of the Donets, Dnieper, and L'vov-Volyn Basins, and in the Northwest where coal supplies are obtained from the Pechora Basin. These modeled flows demonstrate the widening supply area of eastern coal as European coal production declines or stagnates (see Mitrofanov, 1977).

This problem is illustrated by some recently published data on Ukrainian coal shipments (*Ugol' Ukrainy*, no.7 (1985), pp.23-24). In 1975, nearly 59 million tons of coal were exported from the Ukraine, or about 27 percent of production. In 1982, exports were still 27 percent of production, but had been reduced by 7 million tons, to about 52 million tons. At the same time, coal shipments *into* the Ukraine increased from 6 to over 9 million tons. About 3 million tons of coking coal were imported from Poland's Silesian field in 1982, with the rest coming from the Kuznetsk Basin (1.2 million tons), Karaganda (1.2 million tons), Pechora (1.7 million tons), and the eastern wing of the Donets Basin in the RSFSR (2 million tons).

Flows of coal are also generated from Karaganda and along the Turk-Sib line from the Kuzbas to southern Kazakhstan, while Central Asian coal is shipped north considerable distances to Aktyubinsk and Uralsk in northern Kazakhstan and through Astrakhan to the European USSR. Some Central Asian coal is also shown as being shipped through Krasnovodsk across the Caspian to supply the Caucasus, although Donets coal has traditionally supplied the Caucasus (Danilov *et al*., 1983, p.428). In contrast, Urals and Georgian coal is strictly limited to local use.

The model generates a small flow of coal from the Far East over the Trans-Siberian line to East Siberia, and then a larger flow from East Siberia to the Kuzbas in West Siberia. This "frees" more of Kuzbas production for shipment west. Shipments of coal from the Far East to East Siberia are apparently incorrect (see Rodgers, 1983, p.196). In 1975, coal production in the Far East was about 35 million tons (Dienes and Shabad, 1979, p.111), and deliveries to the region were 36 million tons (Table 6.6), implying that regional exports are nil. However, East Siberian coal is known to be imported into West Siberia (Kontorovich, 1982, p.vi; Lydolph, 1979, p.285; Kazanskiy, 1980, p.38), although Kuznetsk coking coal is shipped in the reverse direction to the Far East for export. About 70 million tons of coal were produced in East Siberia in 1980, but only 80 percent was consumed within the region (Danilov et al., 1983, p.336). Presumably, the remainder (about 14 million tons) is sent to West Siberia. For example, part of the needs of the Novosibirsk TETs is met with Kansk-Achinsk coal (P.K. Dubinskiy, 1977, p.23).

In general, the model evidently replicates the broad dimensions of Soviet coal movements and general supply patterns (Lydolph, 1979, pp.285, 429; Kazanskiy, 1980, pp.36-40), although there are some discrepancies (e.g., the Far East, Caucasus). In particular, the model correctly shows the massive flow of coal from the eastern basins to the European USSR, right into the Donets Basin itself. Interestingly, Angren lignite from Central Asia is shown as travelling considerable distances, although not as far as Ryazan', where some of it is actually shipped.

However, the overall magnitude of the model-generated flows is too high. The model generates 795.2 billion ton-kilometers, whereas in actuality, only 598.8 billion ton-kilometers were incurred on Soviet railroads in 1980 (Narkhoz SSSR 1980, p.295), and this also includes movements of coking coal. The minor amounts of river and sea shipments cannot explain the discrepancy. The higher volume of traffic in the model may result from overestimating coal consumption because losses are not taken into account. Thus, more raw coal may be being shipped in the model than is actually the case. For example, the model generates a flow of 137.5 million tons of coal to the European USSR although published data indicate the flow is only around 120 million tons, and this includes shipments of coking coal.

A very rough estimate of railroad freight traffic for coking coal would be 146 billion ton-kilometers; i.e., coal's average length of haul (818 kilometers) multiplied by production (178 million tons). Thus steam coal freight on railroads may generate approximately 453 billion ton-kilometers in 1980, which with sea and river movements would sum to around 473 billion ton-kilometers.

This level of freight turnover (474.2 billion ton-kilometers) is generated in a model simulation with the overall level of demand reduced by 30 percent, from 538 million tons to 377 million tons. This would indicate that losses of steam coal (the difference between gross production and actual deliveries) was around 161 million tons in 1980, or about 30 percent, a significant amount. Tretyakova (1980) calculated that in 1972 losses during processing came to about 28 percent, but deliveries to consumers were much higher than gross output minus losses, implying the greater use of coal by-products in the national economy. Since deliveries to central electrical stations were 299 million tons (Table 6.3), this means that deliveries to other users would be only 78 million tons or so, or about 14 percent of the gross output of non-coking coal. In 1972, these other users accounted for about 19 percent of coal consumption (Kurtzweg and Tretyakova, 1982, p.365). Thus this estimate of 78 million tons appears far too low.

Deliveries are underestimated in this second simulation because it generates an average length of haul of 1258 kilometers to provide a total turnover of 474 million ton-kilometers. Although this is shorter than in the first simulation (1478 kilometers), it is still 54 percent greater than the actual distance (818 kilometers). Therefore, in addition to the effect of losses, the model is generating too much traffic because the coal is being shipped too far.

This is the result of inaccuracies in estimating the distribution of steam coal consumption by the other users. This implies that the consumption of steam coal by industrial, municipal, and household boilers does not reflect the sector's general spatial distribution, but is more concentrated in the coal-producing areas. Thus, the fuel needs of these users located elsewhere are met mostly with alternate, more transportable fuels. A key factor is the 20 million tons produced in unknown basins, which was assigned to Mezhdurechensk. This coal is produced in a variety of small fields, entirely for local consumption. But located at Mezhdurechensk, it becomes part of the long-distance flow to the European USSR. In retrospect, the amount should have been removed from the analysis, and much greater care should have been used in working out regional coal consumption before allocating it to individual nodes.

Flows for this second simulation are shown in Figure 6.3. The largest flow, 62.6 million tons, is from Tselinograd to Kokchetav. The flow from Ekibastuz to Tselinograd is 56.2 million tons, or about 2800 carloads per day, closely approximating the actual amount (see above). The flow of Kuznetsk coal westward is 49 million tons. Thus the model generates an east-west flow of 96 million tons of steam coal, only 20 million tons more than the estimated amount (Table 6.6).

The capacity of only a single supply node, Ekibastuz, needed to be increased to obtain a feasible solution in the first simulation (i.e., that using 100 percent of apparent consumption). Supply at Ekibastuz had to be increased to 85 million tons, although production in 1980 was only 67 million tons (Table 6.2). This adjustment implies that further expansion is needed at Ekibastuz and also serves to justify its rapid development in the last decade. In the second simulation, with only 70 percent of apparent consumption, 59.5 million tons of Ekibastuz coal were utilized, an amount close to actual deliveries (62.4 million tons).

The model also generates a set of opportunity costs for the supply nodes, indicating how much transport cost could be decreased per ton of extra capacity. Not surprisingly the nodes with the highest opportunity costs (around 8.5 million ton-kilometers each) are all in the European USSR. These include the Dnieper Basin, Donets Basin (Pavlograd), and L'vov-Volyn Basin, with somewhat lower opportunity costs (around 8 million ton-kilometers) for the other Donets nodes (Gukovo, Donetsk, Voroshilovgrad, Novoshakhtinsk) and the

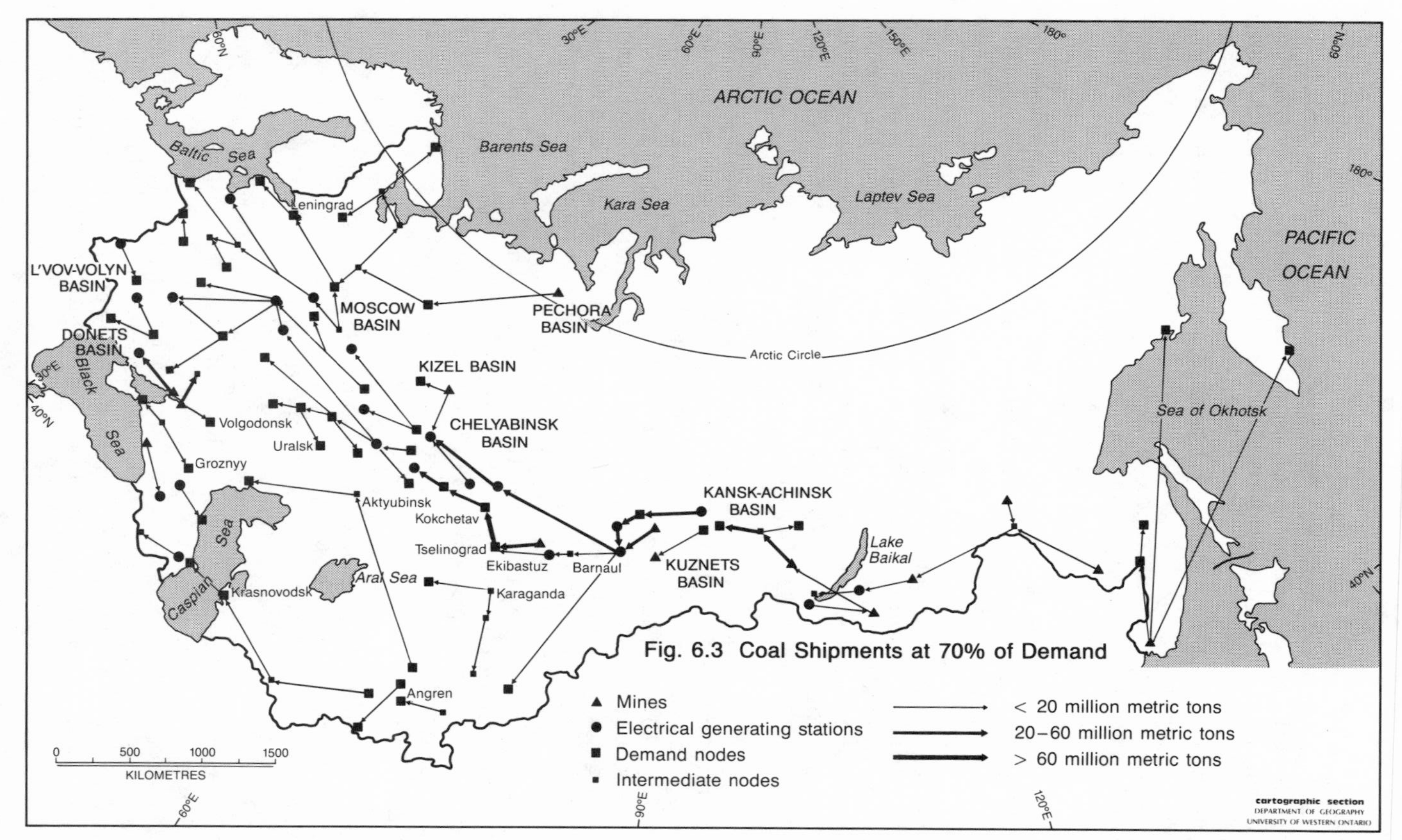

Fig. 6.3 Coal Shipments at 70% of Demand

Moscow Basin. Since the average cost for transporting coal is about 1980 rubles per thousand ton-kilometers (Shafirkin, 1978, pp.222-223), these savings translate into about 1.6 million rubles annually for every extra ton produced at these locations. This illustrates why the Coal Ministry has attempted to maintain production in the Donets Basin despite increasingly unfavorable mining conditions and high costs (Tretyakova, 1985). In contrast, the lowest opportunity costs are in the Far East and East Siberia, reflecting their isolation from the main consuming centers.

D. CONCLUSIONS

The modeled flows seem to be a generally accurate simulation of actual patterns of flows, after taking into account losses, coal heterogeneity, and the crude nature of some of the consumption estimates. This would imply that the distribution of coal is fairly efficient in the USSR.

The massive flows of coal from the east are of magnitudes that must be straining railroad facilities to the limit. Although the model indicates further expansion at Ekibastuz appears warranted, any further expansion will require more transport capacity. A shortage of rail cars was already constraining output at Ekibastuz in 1980 (Nekrasov and Troitskiy, 1981, p.228). Because the railroad cannot handle any additional freight, it is planned for further increments in mined coal to be consumed at large electrical power stations at Ekibastuz, with electricity being transmitted to consumers in the Urals and Central Russia over long-distance, extra-high-voltage lines (Nekrasov and Troitskiy, 1981; "News Notes", SG-April 1984, p.282; March 1980, pp.187-190); i.e., "coal by wire". The first Ekibastuz station reached its designed capacity of 4000 MW in 1984 ("News Notes", SG-April, 1985, p.303), while work on the second Ekibastuz station is underway. The extra-high-voltage lines designed to carry the electricity from Ekibastuz are under construction, with one line reaching as far as Kokchetav ("News Notes", SG-April 1984, pp.290-291).

Despite its considerable potential for expansion, output in the Kuznetsk Basin actually declined in the late 1970s and has been stagnant in the early 1980s, partly because of the strain expansion would place on the railroads. As a result, the Coal Ministry has attempted to maintain production in the Donets Basin despite increasingly unfavorable mining conditions. However, the continuing stagnation of output in the Donbas has led some planners to advocate redirecting investment to the Kuzbas. Alternative transport modes have been proposed to handle the increased freight traffic which would result from such a strategy, such as a long-distance coal-slurry pipeline (Campbell, 1980; Office of Technology Assessment, 1981, pp.98-99), as well as various improvements to the rail system, such as the reconstruction of the Central Siberian line into a specialized coal carrier (CIA, 1980; "News Notes", SG-April 1984, p.280; Kontorovich, 1982).

Development plans for the Kansk-Achinsk Basin are also for a coal-by-wire approach as at Ekibastuz. However, because it is even further from the sources of demand, its development is being given a lower priority ("News Notes", SG-March 1983, pp.249-256). The first mine-month station in the basin, Berezovskoye, projected to eventually reach 6400 MW, is running behind schedule. The first 800 MW unit was originally scheduled to become operational in 1984, but will probably not get going until 1986 ("News Notes", SG-April 1984, p.291; April, 1985, p.303).

The results from the model both explain and generally justify these plans for the coal and electrical sectors.

VII
Transmission Constraints in Soviet Electricity

A. INTRODUCTION

The spatial dimensions of electrical flows in the Soviet Union for 1980 are analyzed in this chapter. Such an analysis is necessary because of recent changes in the Soviet energy situation since the mid-1970s. Within the electrical energy sector the occurrence of periodic blackouts and brownouts are indications that the availability of electric power is becoming problematic, and the impact on industrial growth is proving to be serious (Schroeder, 1985; Sotsialiticheskaya industriya, October 7, 1985, p.1). The increased levels of consumption since the mid-1970s have led to major revisions in construction plans, an increased emphasis on extra-high-voltage (EHV) technology, and lower reserve capacities and higher load factors.

The necessity to deal with these problems by the world's largest centrally planned economy involves large scale capital investment. Given the scale of such investment, the necessity of proper planning is obvious. An evaluation of the wisdom of announced investment therefore becomes vital. This chapter evaluates Soviet announced intentions and determines where the major constraints on production and distribution of electrical power exist.

Electrical power is analyzed as it is a key component of the total Soviet energy economy and one of its most versatile forms of energy. Electricity generation accounted for nearly 40 percent of the primary energy available for domestic consumption in 1980 (Campbell, 1983, p.198), whereas in 1975, it was less than 30 percent (Dienes and Shabad, 1979, p.19). This reflects the fact that an increasing proportion of energy has been consumed in the form of electrical power as the Soviet economy has developed. This pattern is typical of many countries because the versatility and technical advantages of electricity in a widening range of manufacturing activities and other uses make electrification an important aspect of the process of economic development. The electrical power industry occupies a central position in Soviet thinking on economic development, summarized in Lenin's famous dictum: "Communism is Soviet power plus the electrification of the entire country".

The transmission of electricity is important because the increasing shipments of eastern coal over the long distances to the European USSR are placing a growing burden on the railroads (see Chapter VI). To resolve this problem, the Soviets plan to ship energy from the eastern coal basins in the form of electrical power from mine-mouth stations by extra-high-voltage (EHV) lines (Nekrasov and Troitskiy, 1981, pp.31-33; Venikov, 1982; CIA, 1985, p.55). Construction on some EHV lines began in the 11th Five-Year Plan as part of this program ("News Notes", _SG_-March 1980, pp.187-190).

Analysis is limited to examining hypothetical patterns of flow under average (assumed to be typical) conditions because of lack of data. Little is known about the magnitudes or directions of actual flows, even among the major grid systems. In order to analyze these hypothetical flows, the Soviet electrical system is generalized into an abstract network consisting of supply nodes (power stations), demand nodes (cities and export points), and bounded arcs (transmission lines).

B. ABSTRACTING THE ELECTRICAL SYSTEM INTO A NETWORK

Only the portions of the system connected into the unified countrywide grid in 1980 are included in the analysis. This comprises nine of the USSR's eleven major grid systems. These are the Central, Northwest, Volga, South, Urals, North Caucasus, Transcaucasus, North Kazakhstan, and Siberian grids (Figure 7.1). The Central Asian and Far Eastern grids are excluded. These large unified grid

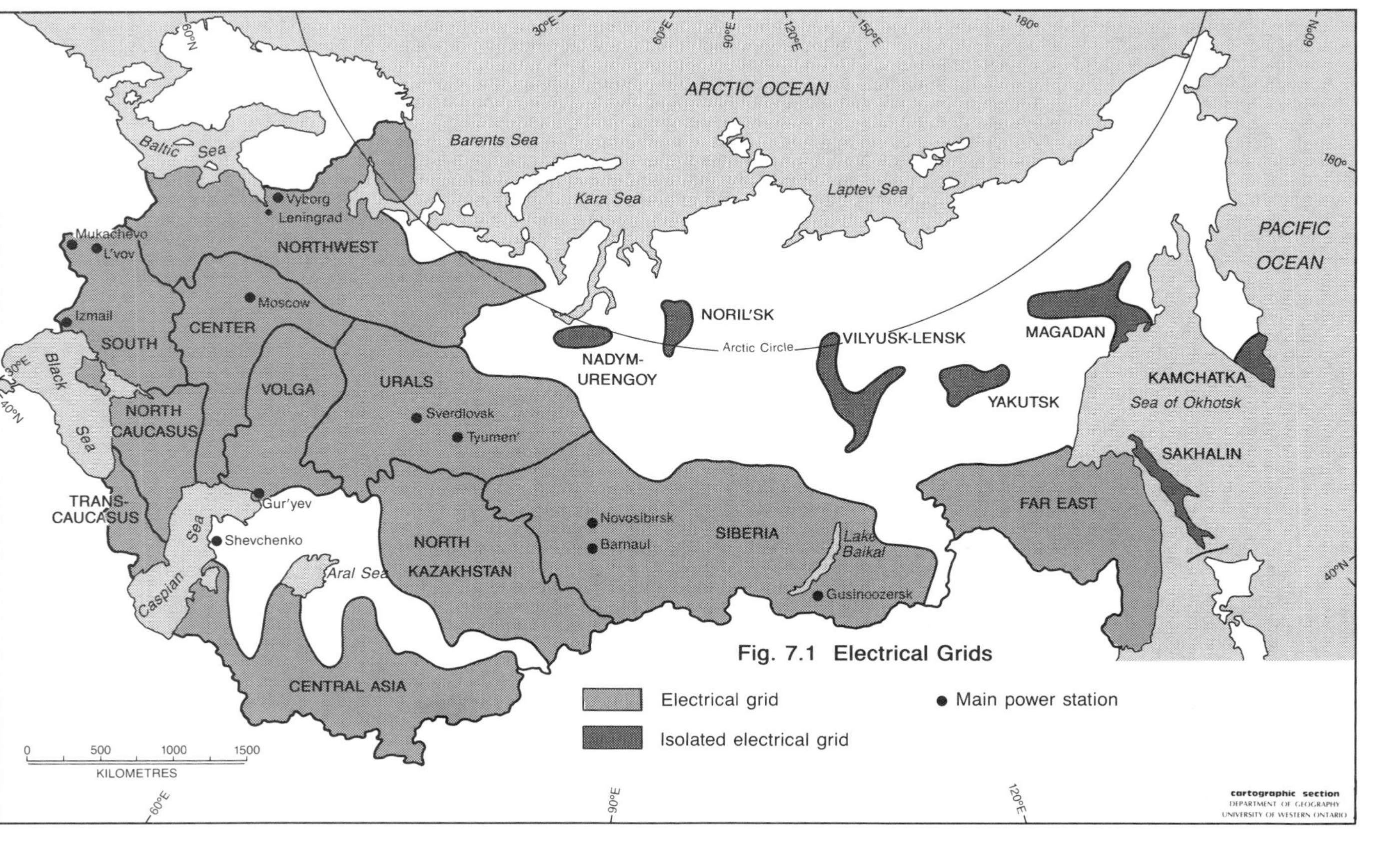

Fig. 7.1 Electrical Grids

systems were formed to take advantage of scale economies, of savings on reserve capacity, and of an expanded choice of generating sources such as shale or hydroelectrical power that have low transportability. Some of the characteristics of the major grids in 1980 are given in Table 7.1.

Supply nodes in the network consist of large Soviet power stations. All stations with capacities in excess of 1000 MW in 1980 are included, as well as selected smaller stations, for a total of 110 supply nodes. The supply available at each is determined by installed capacity, subject to the limitation that the total for each grid equals its average availability (its capacity factor multiplied by total capacity; both are given in Table 7.1); i.e., approximately the grid's average annual load.

For example, the large stations included in the analysis for the Northwest grid (Baltic, Kirishi, Leningrad, Lithuanian, Kola, Bereza, Estonian, Plyavinas, Lukoml', Pechora, Cherepovets, and Riga) have a total installed capacity of 17,317 MW. The average availability for the Northwest grid in 1980 is only 14,317 MW (.57 x 25,900) (Table 7.1). Thus the supply available at each station is set at 85.3 percent of installed capacity. This particular percentage varies by grid system. Small stations make up a much larger part of installed capacity in the Urals and Volga grids, so the larger stations considered in the analysis must be included at a higher percentage of their installed capacity to sum to the average availability for these grid systems.

In the network, demand nodes are used to represent electricity consumption sites. While data on the sectoral distribution of electricity consumption are available in some detail, data on the spatial distribution of consumption are fairly limited. Consumption data are reported for the fifteen republics (Table 7.2), but not for any smaller areas. However, some spatial disaggregation of consumption is possible for the RSFSR and Kazakhstan, two of the largest republics.

The northeastern part of Kazakhstan is in the North Kazakhstan grid and the Gur'yev area in western Kazakhstan was connected with the Volga grid

TABLE 7.1

Characteristics of the Soviet Grid Systems in 1980

System	(1) Installed Capacity (MW)	(2) Percent Hydro	(3) Annual Production (billion KWH)	(4) Average Annual Load (MW/hr)	(5) Capacity Factor
USSR Unified Grid	223,400	19.4	1152.7	131,600	.59
of which:					
Center	39,300	9.4	197.1	22,500	.57
Volga	15,700	26.1	77.8	8,900	.57
Urals	30,800	5.5	188.1	21,500	.70
Northwest	25,900	15.8	130.1	14,900	.57
South	44,600	9.0	249.8	28,500	.64
N. Caucasus	10,000	19.0	49.0	5,600	.56
Transcaucasus	10,500	38.1	43.0	4,900	.47
N. Kazakhstan	8,900	12.4	44.4	5,100	.57
Siberia	35,100	53.3	162.7	18,600	.53
Other	2,600	0.0	10.7	1,200	.47
Central Asia	18,400	35.3	71.7	8,200	.44
Far East	4,000	32.5	15.5	1,800	.44
All Integrated Grids	245,800	20.8	1239.9	141,500	.58
Isolated Grids	20,900	5.7	54.0	6,200	.29
USSR TOTAL	266,700	19.6	1293.9	147,700	.55

Source: Columns 1-3, Nekrasov and Troitskiy (1981, p.200); column 4, yearly production (column 3) divided by 8760; column 5, average annual load (column 4) divided by maximum possible load (installed capacity (column 1) multiplied by 8760.

TABLE 7.2

Republic Electricity Consumption in 1980
(billion KWH)

Republic	Production	Useful Consumption	Losses	Total Consumption
RSFSR	804.9	752.2	63.7	815.9
Ukraine	235.9	199.9	19.2	218.4
Belorussia	34.1	29.1	2.9	32.0
Kazakhstan	61.5	66.8	5.3	72.1
Georgia	14.7	11.7	2.3	13.9
Azerbaijan	15.0	15.0	2.1	17.1
Lithuania	11.7	10.1	1.4	11.6
Moldavia	15.6	7.3	1.2	8.4
Latvia	4.7	7.3	1.2	8.2
Kirgizia	9.2	5.6	.6	6.2
Tadjikistan	13.6	8.8	.9	9.7
Armenia	12.0	10.4	1.4	11.8
Turkmenia	6.7	4.9	.8	5.7
Estonia	18.9	7.2	1.0	8.2
Uzbekistan	35.4	33.3	3.0	36.3
USSR TOTAL	1293.9	1082.9	191.1	1274.0

Sources: Electro-balances of the various republics published in the republic statistical handbooks; Armenia determined as a residual; USSR totals from Nekrasov and Troitskiy (1981, pp.45-71).

between 1976 and 1980 (Nekrasov and Troitskiy, 1981, p.198), so both areas are part of the unified countrywide system. The southern part of the republic is in the Central Asian grid, and a small isolated system exists in the Shevchenko area (Figure 7.1). The latter are not part of the unified countrywide system, and thus are not included in the analysis. Consumption for the different parts of Kazakhstan can be estimated using the information in Tables 7.1 and 7.2. Apparently 13.8 billion KWH were consumed in the Kazakhstan portion of the Central Asian grid and 8.8 billion KWH were produced (and therefore probably consumed) at Shevchenko and Gur'yev. Imports or exports cannot be ruled out for Gur'yev, although they are unlikely. Gur'yev's integration into the Volga grid was probably for flexibility and exchanges of power. The 8.8 billion KWH is allocated between the two areas based upon the size of the cities. Consumption at Shevchenko is estimated at 4.1 billion KWH, with 4.7 billion KWH at Gur'yev. Thus the remainder (49.5 billion KWH) was consumed in the North Kazakhstan grid, implying that it was a net importer of about 5.1 billion KWH in 1980 despite the ongoing development of the Ekibastuz power complex.

Within the RSFSR, consumption in the Far Eastern grid was 15.5 billion KWH and 49.9 billion KWH were consumed in its isolated systems (Table 7.1). Since the RSFSR's isolated systems in 1980 were all in the east (e.g., Noril'sk, Nadym-Urengoy, Vilyusk-Lensk, Yakutsk, Magadan, Sakhalin, Kamchatka) (Figure 7.1), consumption in the Far Eastern and the other isolated grids (15.5 and 49.9 billion KWH, respectively) can be subtracted from the RSFSR total (815.9 billion KWH) to arrive at the amount consumed in the portion of the RSFSR within the national unified grid (i.e., 750.5 billion KWH).

Ryl'skiy (1981, p.108) infers that electricity consumption in the RSFSR's eastern regions in 1980 was about 225 billion KWH, of which 15.5 billion would be in the Far Eastern grid and 49.9 billion KWH in the isolated systems. Thus consumption was about 160 billion KWH in the Siberian grid and the portion of Tyumen' Oblast in the Urals grid in 1980. Thus, 590 billion KWH were consumed in the European portion of the RSFSR, with 20 percent (119 billion KWH) in the Northwest and Volga-Vyatka regions and 70 percent (414 billion KWH) within the Center, Volga, and Urals (Ryl'skiy, 1981, p.101), leaving 10 percent (59 billion KWH) for the North Caucasus and Central Chernozem regions.

Since the republics are generally too large to serve as demand areas, sub-republic (oblast-level) units are used, with their administrative centers serving as the demand nodes in the network. Average hourly consumption in each region is allocated among its oblast-level units according to population, with two exceptions. This includes the other nine republics in the national unified grid (Ukraine, Belorussia, Georgia, Azerbaijan, Lithuania, Moldavia, Latvia, Armenia, and Estonia) plus consumption as estimated above for the regions of the RSFSR and Kazakhstan. Each region or republic's gross consumption in billion KWH is divided by 8760, the number of hours in a year, to provide the average hourly consumption in each area. The allocation within the Ukraine uses 1967 data from Voloboy and Popovkin (1972, p.221) on relative levels of consumption among the Ukrainian oblasts to distribute the Ukrainian total for 1980. Electricity consumption in the Urals region amounts to about one-sixth of Soviet production (Danilov et al., 1983, p.311), or about 216 billion KWH (24,618 MW per hour) in 1980, leaving 198 billion KWH to be allocated among the administrative centers of the Central and Volga regions.

The use of population data as a surrogate for electrical power consumption is justified in the Soviet case, although industry is the major user (Nekrasov and Troitskiy, 1981, p.46). The spatial distributions of industrial production and population largely coincide (Runova, 1976), and there is a high correlation between population and electricity consumption, at least among the republics (Sagers and Green, 1982, p.294). Also, where per capita consumption is known to vary significantly within the republics, such as within the Ukraine, the allocation is adjusted to reflect this.

Administrative centers serve as the domestic demand nodes. They are representative of the demand within each oblast because the industry and urban population tend to be highly concentrated there (Huzinec, 1978). In the few instances where an oblast contains a major industrial center that is not the administrative focus, such as Cherepovets, Magnitogorsk, or Novokuznetsk, the total estimated demand for the oblast is split among the cities according to population size. A total of 148 domestic demand nodes (cities) are included in the network. The largest demand node in the system is Moscow, accounting for about 7.4 percent of Soviet consumption. The second largest is Sverdlovsk, rather than Leningrad, a reflection of the high level of industrialization of the Urals.

Exports, totalling 19.9 billion KWH in 1980 (Vneshnaya torgovlva SSSR 1980, p.25) (2272 MW per hour), are assigned to five export points at the Soviet border. The first, covering exports to Poland, is located north of L'vov (and is also connected with Brest), while the second, representing exports to Hungary, Czechoslovakia, and Romania, is located at Mukachevo. Exports to Bulgaria are assigned to a node located at Izmail, while exports to Mongolia are covered by a node south of Gusinoozersk. Exports to Finland are represented by a node at Vyborg (Figure 7.1).

In addition to the 148 domestic demand nodes, 5 export nodes, and 110 supply nodes, 9 intermediate nodes are added to allow for junctions or changes in capacity at other than supply or demand nodes. The intermediate nodes are cities or towns (Baykalsk, Bugul'ma, Syzran', Karsun, Mikhaylov, Konosha, Mikun, Chudovo, and Arzamas).

Bounded arcs represent the major electrical transmission lines (Figure 7.2). The capacity of each arc is set by an upper bound determined by the rated kilovolt-amperes (KVA) for the represented lines. The USSR has high-voltage lines of 220, 330, 500, and 750 alternating current (AC) KVA and 800 direct current (DC) KVA. Lines of 1150 KVA are under construction, and there are plans to build 1500 (DC) KVA lines. Although the capacity of these lines varies with construction details and weather conditions, these lines are given capacities of 250, 400, 1000, 2000, 700, 3400 and 6000 MW respectively (Venikov, 1982, p.36; Campbell, 1980, p.181).

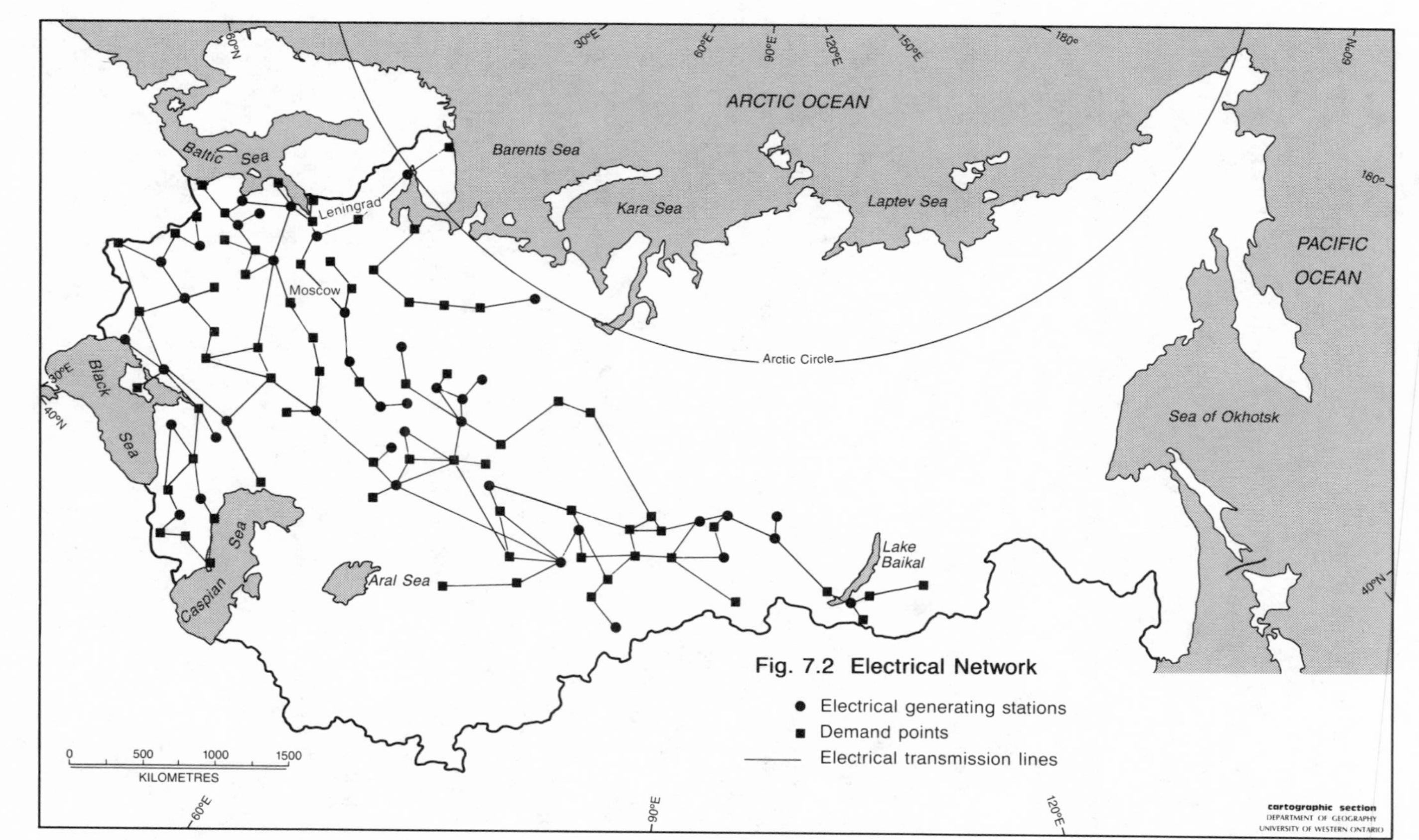

Fig. 7.2 Electrical Network

Although it is planned for the 1150 KVA lines to eventually transmit at 3400 MW, initially the lines are being powered at only 500 KVA (1000 MW) ("News Notes", SG-March 1980, p.190; Kaz. Pravda, December 25, 1983, p.1). The alignment of these lines is taken from several Soviet maps of the system (e.g., Neporozhniy, 1980; JPRS L/10822, 1982).

Distance is a critical consideration in the cost of transporting electricity because of energy loss in the transmission lines as distance increases. A number of factors influences line resistance and therefore power loss. A key factor, however, is voltage; the higher the voltage, the less the line loss. In general, the power loss amounts to approximately 10 percent for every hundred miles (Green and Mitchelson, 1981). Thus, without extra-high-voltage lines, line losses restrict power transmissions to less than 1000 miles (Dienes and Shabad, 1979, p.211). Therefore in the model, the length of an arc represents the cost of utilizing it. The length of each arc is determined as the spherical distance between terminal nodes, calculated from their latitude and longitude.

C. RESULTS

The problem as outlined above turns out to be infeasible because of the widespread lack of transmission capacity to meet the consumption needs of the demand nodes. This result is not totally unexpected, as the system already was capacitated in many areas in 1975 (Sagers and Green, 1982). Although gross consumption of electricity increased only by 24 percent between 1975 and 1980, while the length of 500 and 750 KVA lines increased by 33 percent (Nekrasov and Troitskiy, 1981, p.183), apparently the increase in consumption was not uniform throughout the country. The extra strain on the system indicates that the increase in consumption was confined largely to the energy-short European USSR.

The infeasibility of the problem as modeled results from several factors. First, actual consumption is more dispersed. Smaller cities, towns, hamlets, and villages contribute to the total consumption for each area and are served by a local distribution network not included in the model. Second, in the larger cities, small TETs (heat and power stations) provide electricity on site, reducing the need for electrical transmission over the main lines from the larger central electrical stations. Third, consumption as assigned to the demand nodes includes losses.

To partially compensate for these factors, the initial estimates of consumption are reduced by 10 percent for all domestic demand nodes. This corresponds to the share of agriculture, one of the most dispersed users, in net consumption (Nekrasov and Troitskiy, 1981, p.46). With this adjustment, the problem becomes feasible. This indicates that consumption is more dispersed in 1980 than in 1975. Indicative of this is agriculture's increased share of net consumption.

In addition to this adjustment to demand, several of the extra-high-voltage lines of 1150 KVA currently under construction or planned must be included in order to obtain a feasible solution. Their level of utilization in the model (discussed below) will indicate for which lines construction appears warranted.

The general pattern of hypothetical flows in the Soviet electrical system for 1980 is shown in Figure 7.3. The largest single flow in the network is between Novokuznetsk and Barnual, in Siberia, over one of the EHV lines still under construction. The model-generated flow is 3400 MW per hour over the arc, its assigned capacity. A flow of 3346 MW per hour is generated between Ekibastuz and Kokchetav, and there is a flow of over 3000 MW between Yermak and Ekibastuz (Figure 7.3); both occur over EHV lines. All told, 25 arcs have flows in excess of 1500 MW (Table 7.3), over twice as many as in 1975 (Sagers and Green, 1982, p.298). Of these, 17 have flows of 2000 MW or more. For about half, these massive flows exceed the arc's rated capacity.

The flows shown in Figure 7.3 are aggregated by major power grid into a flow matrix. This matrix indicates an exchange of 13,833 MW per hour among

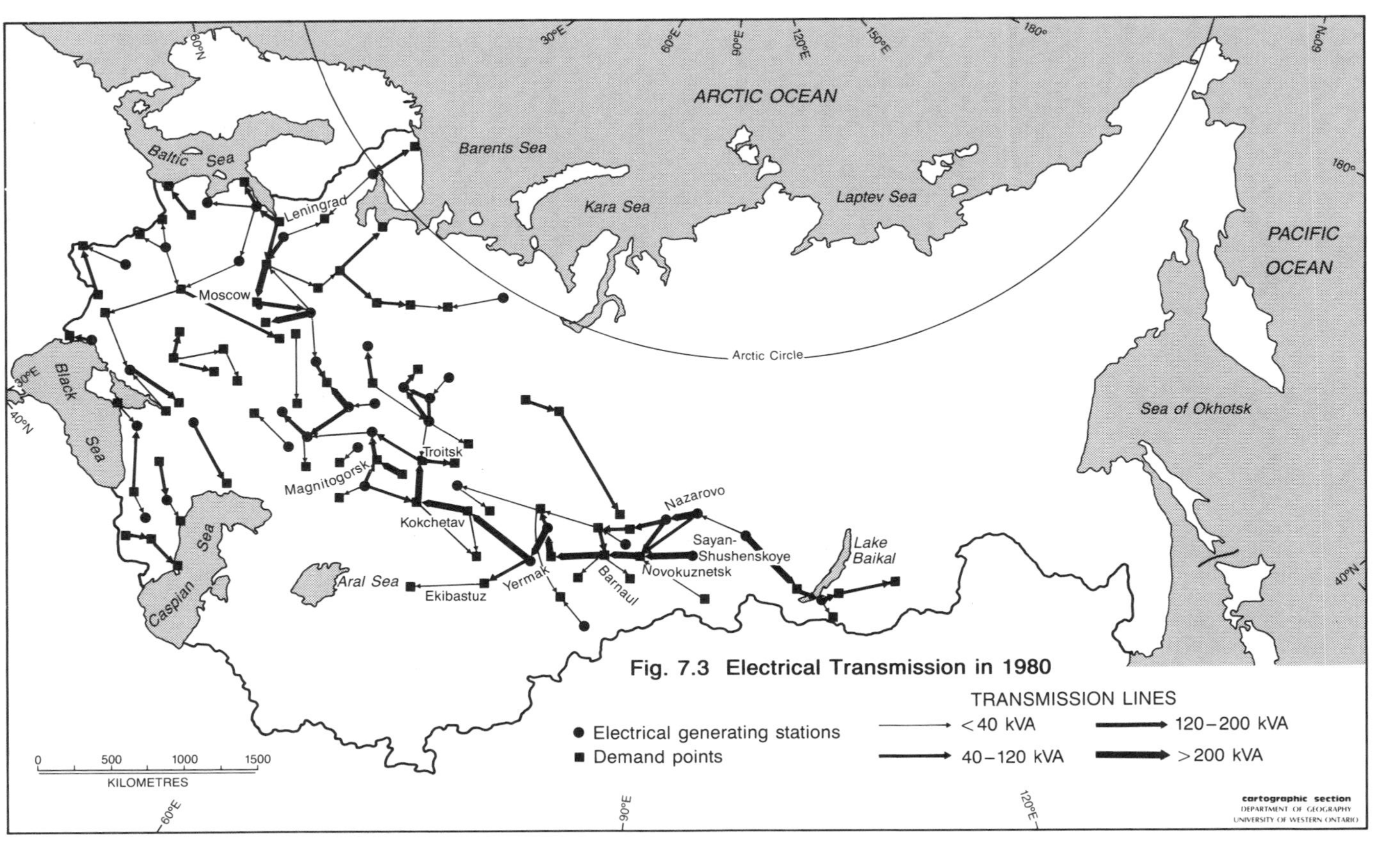

Fig. 7.3 Electrical Transmission in 1980

TABLE 7.3

Power Transmissions Over 1500 MW per Hour

Arc	Flow	Arc Capacity	Grid
Novokuznetsk - Barnaul	3400	(3400)	Siberia
Ekibastuz - Kokchetav	3346	(3400)	Kazakhstan
Yermak - Ekibastuz	3069	1000 + (3400)	Kazakhstan
Kokchetav - Kustanay	2750	(3400)	Kazakhstan
Kustanay - Chelyabinsk	2746	(3400)	N.K.-Urals
Krasnoyarsk - Nazarovo	2614	2000	Siberia
Pavlodar - Yermak	2590	1000 + (3400)	Siberia-N.K.
Bratsk - Irkutsk	2526	2250	Siberia
Konakovo - Moscow	2500	3000	Center
Zainsk - Kazan'	2427	1000	Volga
Barnaul - Pavlodar	2411	(3400)	Siberia-N.K.
Votkinsk - Izhevsk	2159	1000	Urals
Chudovo - Kalinin	2139	2400	NW-Center
Sayanogorsk - Novokuznetsk	2004	(3400)	Siberia
Kashira - Moscow	2000	2000	Center
Troitsk - Magnitogorsk	2000	500	Urals
Reftinsk - Sverdlovsk	2000	1000	Urals
Magnitogorsk - Ufa	1956	1000	Urals
Saratov - Syzran'	1846	1000	Volga
Uglegorsk - Gorlovka	1742	400	South
Gusinoozersk - Ulan-Ude	1738	500	Siberia
Nazarovo - Anzhero-Sudzhensk	1616	2000	Siberia
Nazarovo - Novokuznetsk	1529	(3400)	Siberia
Kirishi - Chudovo	1500	400	Northwest
Sredneuralsk - Sverdlovsk	1500	1000	Urals

Note: Line capacities in parentheses are for the planned lines of 1150 KVA.

the major grid systems, or about 121 billion KWH annually. This amounts to 9.4 percent of Soviet electricity production. Together with Soviet exports to neighboring countries, the modeled inter-regional exchange of power is about 10.9 percent of production. This represents a considerable increase in the proportion being transferred among the grid systems, as it was only 5.4 percent in 1970 and about 7.7 percent in 1975 (Sagers and Green, 1982, p.297). Increased regional interdependence results from the same factors that led to the organization of the large, unified grid systems: scale economies, savings on reserve capacity, and the utilization of generating sources with low transportability.

The flow matrix shows the Center to be by far the largest importer, just as it was in 1975. Its net imports of power amount to 14.4 billion KWH annually (Table 7.4). This is at about the same level as they were in 1975 (Dienes and Shabad, 1979, p.194). Most of the imported power comes from the Northwest grid, with smaller amounts from the Volga grid and the South.

The model indicates that the other major power importer is the Urals grid. The region is in fact reported to import considerable electricity from the neighboring regions (Danilov _et al._, 1983, p.312). The model shows the Urals grid required about 7.2 billion KWH more than 1980 production (Table 7.4).

The Volga grid shows a continuation of the 1970-75 trend, with a slow deterioration in its relative position. It was one of the largest exporters in the early 1970s, and still was a net exporter of several billion KWH in 1975 (Dienes and Shabad, 1979, p.194). However, the model indicates the grid had a slight power deficit by 1980 (Table 7.4).

The South, which had been the largest exporter in 1975, still is shown as being a net exporter in 1980, but only of some 1.9 billion KWH (Table 7.4), most of which goes to the Center. In contrast, it had a surplus of about 20 billion KWH in 1975 (Dienes and Shabad, 1979, p.194). However, the South did account for almost all of the 19.9 billion

TABLE 7.4

Power Transmissions Among the Grid Systems
(MW per hour)

Grid	Exports	Imports	Net	Annual Billion KWH
Center	1500	3143	-1643	-14.4
Volga	402	481	- 79	- .7
Urals	1923	2746	- 823	- 7.2
Northwest	2139	1680	459	4.0
South	804	585	219	1.9
N. Caucasus	879	508	371	3.2
Transcaucasus	486	294	192	1.7
N. Kazakhstan	3829	3253	576	5.0
Siberia	2411	1143	1268	11.1

Source: Data compiled from flows shown in Figure 7.1.

KWH exported abroad in 1980, so most of the grid's power surplus in 1980 evidently was being transmitted to other countries rather than to other regions within the USSR.

Surprisingly, the largest exporter of power in the model is the Siberian grid (11.1 billion KWH) (Table 7.4). This power surplus is transmitted through the planned EHV lines across Kazakhstan to the Urals. The modeled results do not reflect the Siberian grid's true situation, as the transmission capacity to effect this transfer did not exist in 1980. Also, the North Kazakhstan grid is shown as a net exporter of power in the optimal solution (Table 7.4), although when its consumption in 1980 was estimated above, it was actually a net importer. The discrepancy is undoubtedly due to including within the model the 1150 KVA line to the Urals from Ekibastuz, still under construction ("News Notes", _SG_-April, 1984, pp.290-291).

But the model does show the Northwest grid as a major power exporter (Table 7.4). In fact, it is now the grid with the largest amount of excess production, because of the large nuclear stations at Leningrad and the Kola Peninsula (_Sotsialiticheskaya industriya_, January 12,, 1986, p.2). This same article calls for the construction of an EHV line extending from the Northwest grid through the Center to the Ukraine and North Caucasus. Although electricity production in both regions is said to be no longer sufficient, basically the new line is for increased maneuverability. The article also calls for extending the 1150 KVA Ekibastuz-Urals line to the Center for greater maneuverability.

Soviet generating capacity is more than ample to meet demand, as demand is satisfied with almost all plants operating at normal generating capacity as assigned. This is as it should be, since the model is of average rather than peak load. Some stations are in fact underutilized in the model. These are Ust-Ilimsk (in Siberia), which is not drawn upon at all (just as it was not in 1975), and utilization at the Surgut (near Tyumen'), Chernobyl', Tripl'oye, and Zaporozh'ye stations (in the Ukraine) is less than 50 percent of their assigned capacities. The only supply node that required extra capacity in order to satisfy demand in its surrounding area is Novosibirsk, a problem that also existed in the 1975 solution (Sagers and Green, 1982, p.299).

However, the spatial separation of supply and demand nodes together with insufficient transmission capacity can lead to potential shortages of electrical power for certain areas. In the Soviet system, the lack of transmission capacity is far more severe than a shortage of generating capacity. Well over a fourth of the arcs in the system have flows equal to or greater than their assigned capacity, a higher proportion than in 1975 (Sagers and Green, 1982, p.299). Although capacitated arcs occur throughout the USSR, the region with the most severe lack of transmission capacity is the Urals. Some arcs in the region (e.g., Troitsk-Magnitogorsk) must operate at 400 percent of initially assigned capacity, and many operate at 200 percent of capacity.

The lack of transmission capacity in the region is also indicated by the opportunity costs provided by the model for each of the arcs. These show the reduction in costs that would occur if more transmission capacity were available for each arc. The arcs with the highest opportunity costs (and therefore serve as the most binding constraints on the efficient distribution of electricity) are in the Urals. The serious power shortage of the Urals region is discussed in several Soviet articles (e.g., <u>Sovetskaya Rossiya</u>, June 16, 1985, p.2). Another area with high opportunity costs is the Moscow region. Other than these areas, the occurrence of capacitated arcs is isolated, usually involving a short connection between a station and a nearby city.

The region with the most severe shortcomings in transmission capacity in 1975 was West Siberia (Sagers and Green, 1982, p.299), particularly the Barnaul-Novokuznetsk area. This problem does not appear in the 1980 solution because a 500 KVA line was built after 1975 between Barnaul and Novosibirsk, alleviating some of the transmission problems in the region.

The level of utilization of the planned 1150 KVA lines indicates that their construction definitely is warranted. They are all heavily utilized, carrying the largest flows in the system (Table 7.3). Since the demand is for 1980, the need for the new lines is obviously quite critical. This particularly affects the Urals because of the region's large demand. Attempting to achieve a solution without the EHV arcs creates difficulties across a large area extending from Moscow to West Siberia.

In a second simulation, the proposed 1500 KVA (DC) line from Ekibastuz to the European USSR is added to the network. The results, shown in Figure 7.4, indicate that this project would be best deferred to a later time. The initial section of the line from Ekibastuz west to Orsk is not utilized at all. Although the other two sections of the Line (Orsk-Saratov and Saratov-Tambov) carry some current (356 and 1589 MW, respectively), the flow is reversed, going from west to east. This perverse flow is the result of the large power demand in the Urals. In fact, a recent article says the 1500 KVA Ekibastuz-Center line is not needed (Sotsialitichiskaya industriya, January 12, 1986, p.2).

The need for expanded development of the Kansk-Achinsk Basin in the near future also seems unwarranted. Construction of the Nazarovo-Novokuznetsk EHV line which connects the Basin's mine-mouth stations with the Kuzbas industrial complex, is needed since it is relatively heavily utilized (Table 7.3). Currently, the power needs of the Kuzbas seem to be met sufficiently by existing capacity (e.g., Sayan-Shushenskoye, Nazarovo), if they are connected with 1150 KVA lines. However, increases in consumption during the 1980s will probably require more thermal generating capacity, particularly to meet winter demand. This could take place in the Kansk-Achinsk Basin, so the slow construction pace envisioned for the Berezovskoye station and the overall development of the Basin during the next decade appears justified ("News Notes", SG-March 1983, pp.249-256). However, a recent article proposes that a 2500 KVA line be built connecting the Kansk-Achinsk Basin with the European USSR (Sotsialiticheskaya industriya, January 12, 1986, p.2).

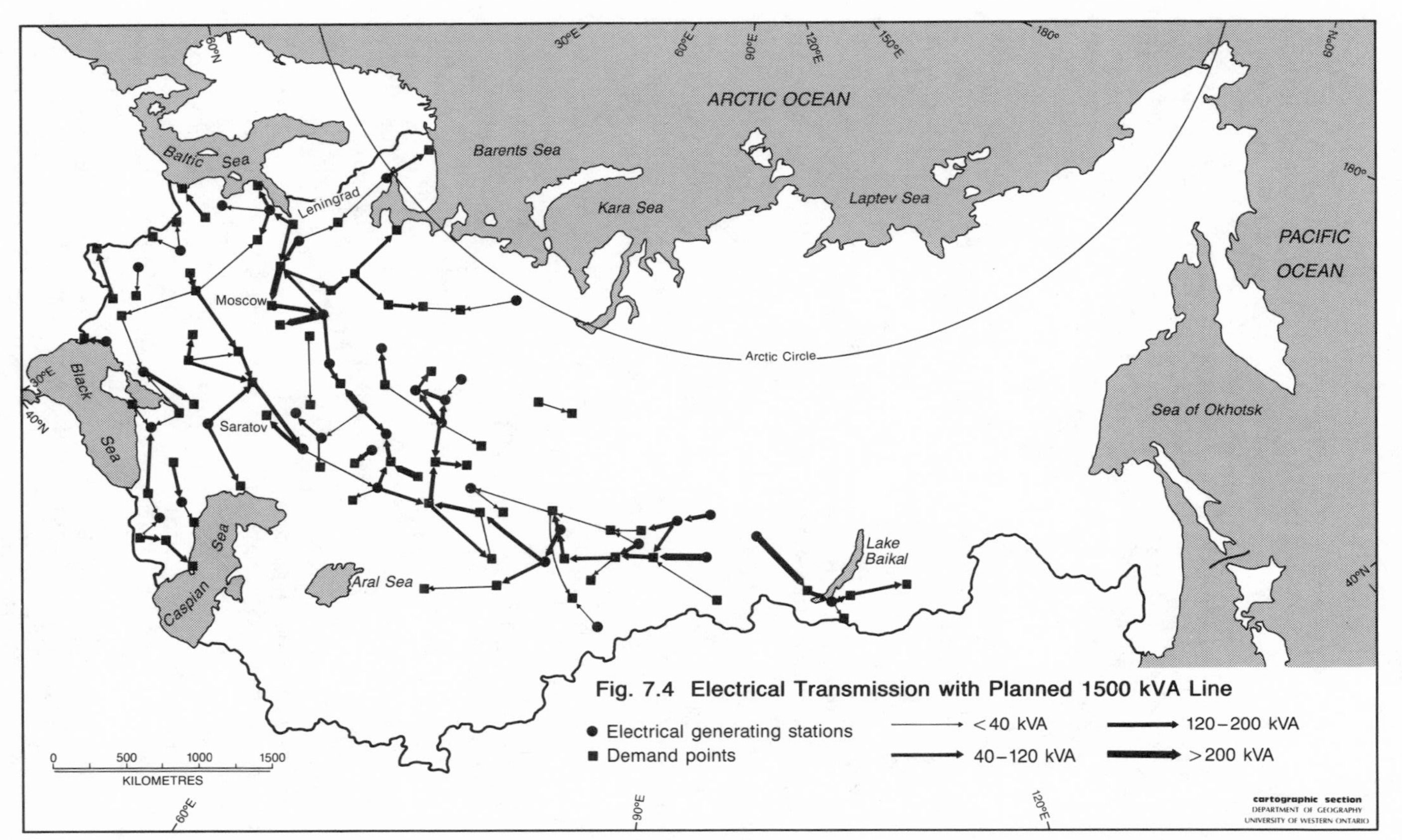

Fig. 7.4 Electrical Transmission with Planned 1500 kVA Line

D. CONCLUSIONS

Overall, the Soviet electrical network is operating under far more strain than in 1975. A higher percentage of the system's arcs are capacitated, and the problem proves to be infeasible without adjusting for greater dispersion of consumption and including planned extra-high-voltage lines as if they were operational. Therefore the analysis is more tentative than for 1975. However, it seems clear that the main problem in the Soviet electrical network is the lack of transmission capacity, particularly in the Urals. The solution gives an approximation of the magnitude and direction of flows of electricity that largely conform with previously established trends.

In terms of policy implications, the planned EHV network of 1150 KVA lines in the Siberia-Kazakhstan-Urals region is definitely needed. All these arcs are heavily utilized in the model. In contrast, the construction of the 1500 KVA (DC) line from Ekibastuz to the Center does not appear justified as yet. This concurs with actual developments. The construciton of the 1150 KVA system has been going forward, whereas work on the 1500 KVA (DC) project ceased in 1981 (CIA, 1985, p.55).

VIII
Summary

Soviet energy resources are the largest in the world, yet energy problems have not bypassed the USSR. A major aspect of current Soviet energy difficulties is the problem of transporting energy. This is because of the USSR's large size and the spatial disparity between energy resources and demand. Transportation has become an increasingly important problem as production has shifted eastward, while consumption has remained concentrated in the western "European" portion. The increasing transport burden has been identified as a major contributor to the poor performance of the Soviet economy in recent years (Schroeder, 1985).

This book analyzes the transportation of Soviet energy resources. The purpose has been to determine the general pattern of movement for each of the main forms of energy (gas, crude petroleum, refined products, coal, and electricity), to identify constraints in the transportation system that inhibit efficient flows, and to evaluate the prospects for future developments, based upon the analysis of the system.

To accomplish this, each energy-transportation system has been modeled as an abstract, capacitated network consisting of supply nodes (production sites), demand nodes (consumption sites), and bounded arcs (transport linkages). The essential characteristics of each are supply availabilities (production), demand levels (consumption), and transport capacity, respectively. The optimal flows and associated costs for each network are determined by applying the out-of-kilter algorithm, a network allocation model, to each abstracted system. This was done for each network using 1980 as the base year.

In general, the models provide justification for the post-1980 developments in each system. Often these developments appear as solutions to the bottlenecks and transport problems identified in the analysis. Thus overall, the current Soviet energy program seems to be a rational response to existing problems. The models indicate that Soviet development plans in the fuels industries are rational, with the distribution and movement patterns approaching optimum efficiency. The major findings and speculations that result from this exercise are summarized below for each of the systems.

Natural gas has the leading role in the Soviet energy program; production increased from 435 to 643 BCM between 1980 and 1985, and plans call for output to reach some 835-850 BCM by 1990. Because increments in gas output are not constrained by reserves, gas production could be increased considerably in a relatively short period of time. The transport constraint for the gas industry, resulting from the location of the reserves in northern Tyumen' Oblast, was lessened by the concentration of the gas in such a small area. Gas flows were modeled for 1970, 1975, 1980, and 1985 because of the considerable changes in the gas system in this period. The modeled flows closely replicate those actually existing in the Soviet system, implying that the Soviet gas system is operating near optimum efficiency. Our models of the Soviet gas network indicate that it became successively more congested between 1970 and 1980, consonant with the maturation of the Soviet gas industry. The major constraint in the pipeline system in 1980 was in the main line from West Siberia, with other major bottlenecks along the Northern Lights route, at the main distribution points of Ostrogozhsk and Novopskov (located south of Moscow), and in the lines to the Transcaucasus. Subsequent pipeline developments in the 11th FYP for the gas network alleviate these constraints, indicating that their construction was warranted.

A scenario for 1985, incorporating five major new lines, while increasing supply and demand at several key nodes, shows an improvement in the overall efficiency of the distribution network, as the most serious constraints in the system are alleviated. This confirms the view that additional pipeline construction was necessary for the success of the gas plans. Also, the rather significant increase in gas exports contemplated between 1980 and 1985, including those to Western Europe, does not particularly tax the gas system with the additional lines installed.

Petroleum is the most transportable of the fossil fuels, so it is not surprising that it was the first of Siberia's energy resources to be developed for widespread use in the European USSR. In the transportation of crude petroleum, the hypothetical flow patterns generated by the model in 1980 generally agree with what little is known about actual flows, implying that the petroleum network also is operating efficiently. The major constraints in the system in 1980 were in the European USSR, particularly in the older oil-producing regions in the Caucasus, rather than in the east at the origin of most of the USSR's petroleum. These problems have been alleviated to a considerable extent by the construction of additional pipelines in the region since, so again the model correctly anticipates subsequent developments. In particular, the new Surgut-Novopolotsk pipeline greatly increases the efficiency of petroleum transmission.

The model also shows the pipelines from West Siberia to the European USSR to be operating close to capacity in 1980. This demonstrates why a new pipeline from West Siberia to the European USSR (between Kholmogory and Klin) was constructed, principally to handle the increment in West Siberian output. Thus, the problems of realignments in supply patterns necessitated by the decline of production in the older producing regions have been largely overcome.

In contrast, the movement of refined products is relatively inefficient. About twice as many ton-kilometers as necessary are generated in distributing refined products in the USSR, primarily because of the mismatch among the various products between local consumption needs and refinery output mix. This implies that Soviet planning is adequate for transporting homogeneous commodities in unique transportation systems (e.g., gas, crude petroleum, electricity), as their distribution is relatively efficient, but does not perform very well with the complexity of heterogeneous products on common carriers. Another problem is that the distribution of refining capacity is a reflection of the past, when different supply and demand factors were operating. A second simulation of refined product flows, incorporating the refineries planned for completion in the 11th Five Year Plan, indicates that these locations are generally justified. Thus, problems in the distribution of refining capacity are being gradually rectified. Beyond this, the model indicates that the most likely areas for future refinery construction are the western Ukraine and the Far East.

The simulation of Soviet coal flows explains and justifies post-1980 plans for the coal sector, mainly in the development of a coal-by-wire approach for the lower quality coal of some eastern coal basins. This is because the model generates a massive movement of coal from the east at volumes that must be straining railroad facilities to the limit. The model also indicates that further expansion of production at Ekibastuz appears warranted. However, this is not possible without more transport capacity. This dilemma appears to be resolved in the Soviet plans to expand production at Ekibastuz, using the fuel to power mine-mouth generating stations and transmit the energy to the Urals in the form of electricity with extra-high-voltage electrical lines. An alternative solution to the transport problem that also is supported by the results of the model is the conversion of the Central Siberian rail line into a specialized coal carrier to transport the higher quality Kuznetsk coal to the European USSR.

The model of the Soviet electrical system indicates that it is operating under far more strain than in 1975. A higher percentage of arcs are capacitated, and the problem proves to be infeasible without adjusting for greater dispersion of consumption and including some planned extra-high-voltage lines as if they were operational. The main problem in the Soviet electrical network is the lack of transmission capacity, particularly in the Urals region. As a result, the model demonstrates the definite need for the planned EHV network of 1150 KVA lines in the Siberia-Kazakhstan-Urals region, since these arcs are heavily utilized in the model. In contrast, a simulation incorporating the planned 1500 KVA (DC) line from Ekibastuz to the Center indicates that the project does not appear justified because of very limited utilization and perverse flow patterns on some of the added arcs.

Therefore, subject to limitations of the data and the models, the Soviet energy program as carried out during the 11th FYP appears to be a response to the problems existing in their energy system in 1980, as well as the opportunities available. Given the transport constraints identified in the flow models, current developments usually appear as ways to alleviate these obstacles or bottlenecks.

In terms of areas for future research and possible improvements to the study, a number of suggestions are offered. One area that needs improvement is in transport costs. Our models incorporated very simplistic cost functions; i.e., cost varied linearly with distance. Another major shortcoming of this study is the lack of appropriate capacities for arcs representing railroads. The problem is very important and needs to be addressed, but even conceptual aspects of the problem are subject to question. For example, judging from the press, the availability of cars appears to be the major constraint in railroad transportation, rather than line or other facilities such as motive power and terminals (Hunter and Kaple, 1983). However, the situation is very murky, particularly given the focus of this study upon individual stretches of line rather than in the aggregate. It would also be helpful in the future to pay more attention to the installation of compressor and pumping stations on the gas and oil pipelines to better account for pipeline capacities along specific arcs.

Another area in which the study could be improved would be a simultaneous solution for the distribution of crude petroleum and refined products, rather than separately. This could be done by linking the two networks into a single model, an approach which would be far more realistic than that taken here.

Finally, the estimates on the regional distribution of coal consumption by other consumers could be improved by working from regional estimates of total coal deliveries for disaggregation to individual nodes. The method utilized for distributing consumption among the nodes turned out to be a major source of inaccuracy in the model.

However, overall the models provide some very interesting insights on the transportation of Soviet energy resources.

IX
Epilogue: The Impact of Chernobyl'

The world's most serious nuclear power accident to date occurred at the Chernobyl' nuclear power plant in April and May, 1986. Apparently, a non-nuclear explosion occurred on April 25, in one of the four operational reactors at the plant. The explosion set off an intense fire in the graphite and nuclear fuel core of the reactor leading to the shut-down of the other three reactors. In addition to the immediate loss of Chernobyl's generating capacity, which will have several consequences in the short term, the accident also is likely to retard the long term development of the Soviet nuclear power industry, which will in turn affect the transportation of Soviet energy resources. This is because it will increase the need for more electric power in the European USSR from conventional thermal plants using coal, natural gas, and oil.

A. SHORT TERM EFFECTS

The Chernobyl' nuclear power station had four operational reactors, rated at 1000 MW each. A fifth was scheduled to start in 1986 ("News Notes," SG-April, 1986). As a result of the accident at reactor no. 4, the Chernobyl' station no longer produces electrical power. This removes 4000 MW of installed capacity from the Southern grid system, one of the USSR's 11 large integrated regional power grids, 9 of which are interconnected to form the national unified network (see Chapter VII). The loss of such a large station in this particular area is likely to have several immediate impacts:

1) increased use of fossil fuels; Chernobyl' generated energy is equivalent to 9.5 million tons of standard fuel a year.

2) decreased energy exports to Eastern Europe; in 1985 Chernobyl' produced more power (29 billion KWH) than all of Soviet electricity exports (26.3 billion KWH); Hungary will be the hardest hit.

3) increased shortage of electricity in the Ukraine. The Ukraine was already experiencing a shortage of electricity before the accident. But the shortage is not likely to be felt until the late fall or winter when the peak demand for electricity occurs.

4) Chernobyl'-type reactors represented 49.5 percent of all Soviet nuclear capacity as of January 1, 1986, and supply power to key industrial areas, including Moscow and Leningrad. If they all were closed because of structural flaws, the economic consequences for the USSR would be very severe.

1. Increased Use of Fossil Fuels

In 1985, the Chernobyl' station generated 29 billion KWH (Summary of World Broadcasts, February 28, 1986, p. 7). Since Soviet thermal electric stations required an average of 326 grams of standard fuel per KWH in 1985 (*Teploenergetika*, no. 2 [1986], p. 3), the operation of the Chernobyl' station generated primary electricity equivalent to 9.5 million tons of standard fuel. This is equivalent to 9.5 million tons of good quality coal, 6.9 million tons of mazut (fuel oil), or 8.1 BCM of natural gas. Thus, one impact of the accident probably will be the increased use of these fossil fuels for electricity generation in the region. And because of the depletion of fuel reserves in the European portion, this extra amount will have to be transported from Siberia.

The purpose of the Soviet nuclear power program is to reduce the use of fossil fuels in the European portion of the USSR and limit the need for transporting fuels thousands of kilometers from Siberia. The Soviet government had planned to stablize fossil fuel-generated electricity in the USSR between 1981 and 1985, mostly through rapid growth of nuclear power in the European portion. Although this goal was not achieved, some progress was made. By the end of 1985, the USSR had 14 operational stations with one or more reactors; nuclear power accounted for 11 percent of all electricity generated, compared with 5.6 percent in 1980 ("News Notes," SG-April, 1986). Nearly all the nuclear stations are in the European USSR, so about 9.3 percent of electricity generated there was nuclear in 1980, and the plan was to raise this to 17.2 percent by 1985. At the end of 1985, the USSR had 28,310 MW of installed nuclear capacity, up from 12,490 MW in 1980.

2. Decreased Electricity Exports to Eastern Europe

The Chernobyl' station was an integral part of the Southern grid system, which supplies nearly all of Soviet electricity exports, most of which go to Eastern Europe. The exceptions are exports to Finland and some Polish exports, which are supplied by the Northwest grid, and the small amount of exports going to Mongolia and Turkey, which are supplied from stations in East Siberia and Armenia. In 1985, Soviet exports were about 26.3 billion KWH, an increase from 1980 when they were 19.1 billion (PlanEcon Data Bank, 1986, p. S-20). Of the 26.3 billion KWH, about 21.1 billion were exported to Eastern Europe (PlanEcon Report, nos. 19-20 [May 16, 1986], p. 9). Electricity production at Chernobyl' exceeded Soviet exports, so it is very likely that the accident at Chernobyl' will result in reduced shipments of electricity to Eastern Europe. Some of the Eastern European countries are quite dependent upon Soviet electricity exports, particularly Hungary, where the 9.4 billion KWH received from the Soviet Union in 1985 represented about 25 percent of domestic consumption (PlanEcon Report, nos. 19-20 [May 16, 1986], p. 9).

Chernobyl' was one of several Ukrainian stations supplying power to Eastern Europe. Chernobyl's first reactor started up in 1977, and exports to Eastern Europe began around 1980, when an extra-high-voltage line, of 750 KVA, was completed, linking Chernobyl' with the terminus of the existing 750 KVA export line at Mukachevo, near the East European border, via the new Rovno nuclear station. This existing 750 KVA export line runs nearly the length of the Ukraine, from the Donbas through Vinnitsa to the western border, and was completed in

1978. In 1982, a second 750 KVA line was completed, from Chernobyl' to Vinnitsa, providing a second connection with the existing export line. These ties to Eastern Europe were further strengthened with the completion of a second Ukrainian export line between the Khmelnitskiy nuclear station (which is scheduled to begin operating in 1986) and Poland in 1984-85, as this also includes an eastward extension from Khmelnitskiy to Chernobyl'. A third export line through the Ukraine was finished in early 1986, but further south, between the new South Ukrainian nuclear station and the Rumanian-Bulgarian border. Another 750 KVA line from Chernobyl' was under construction in 1986, going eastward from the station to the neighboring Kursk grid system.

3. Increased Shortage of Electricity in the Ukraine

The amount of electricity shipped from the Southern grid to Eastern Europe now represents nearly all its surplus electricity. In the mid-1970's, the Southern grid supplied not only exports, but also about 20 billion KWH to the neighboring regions of the USSR, mainy the Central grid system (Sagers and Green, 1982, p. 298). But this surplus no longer exists, and in fact, the Ukraine now is beginning to experience a shortage of electricity for the first time (Chapter VII; *Sotsialiticheskaya industriya*, January 12, 1986). Also, the Southern grid operates at a relatively high capacity factor (.64 in 1980), one of the highest in the USSR; only that for the Urals grid is higher. The result of this is that there is relatively little reserve capacity in relation to peak load, so the loss of the Chernobyl' station will have a significant impact on the availability of electricity, over a year. This is particularly important to consider in light of the extreme tightness we discovered in the Soviet electrical system.

The time element is important because during a relatively short period, other stations in the grid can be operated closer to maximum capacity to cover the shortfall. But over a longer period, this option is untenable because of regular downtime at the other stations caused by the need for periodic maintenance as well as equipment failure and technical difficulties.

Another factor to consider in assessing the immediate impact is that the peak load in the Soviet system occurs in the winter, so electricity consumption now is much less than generating capacity, even without Chernobyl'. Soviet electricity production in June is typically only 70-75 percent that of December and January.

4. The Consequences If All Chernobyl'-Type Plants Were Closed

An unconfirmed report, which would have a significant impact if true, was that all other reactors of this type would be temporarily shut down. The light-water-cooled, graphite-moderated reactors employed at Chernobyl', known by Soviet designation RBMK, are one of two widely used types in the Soviet nuclear program. At the end of 1985, the RBMK-1000 comprised 14,000 MW, or 49.5 percent of all installed nuclear capacity; together with two older RBMK-type reactors at Beloyarskiy, near Sverdlovsk, of 100 MW and 200 MW, and the new RBMK-1500 installed at Ignalina in 1983, these type of reactors comprised 55.9 percent of all Soviet nuclear capacity.

The stations which use the same type of reactors as Chernobyl' are Leningrad, which has 4 reactors, Kursk, which also has four, and Smolensk, which has two. If these units were shut down, electricity demand in several key areas of the European USSR could not be met. Leningrad's four units represent a significant share of installed capacity in the Northwest grid, as well as just about the only excess capacity in the entire European USSR (_Sotsialiticheskaya industriya_, January 12, 1986). They also supply electricity exports to Finland over a new high voltage line completed in early 1985. The Kursk and Smolensk stations are critical components of the important Central grid containing Moscow. The Central grid is a large power importer, most of which now comes from the Leningrad nuclear stations (see Chapter VII). The Kursk station also supplies the large ferrous metallurgy complex of the KMA (Kursk Magnetic Anomaly); a 750 KVA line to the new steel plant at Staryy Oskol' was completed in late 1985. The two units at Smolensk supply power to Moscow through a new high-voltage link, finished in late 1985. No other supplies of electricity are available for these important industrial areas, so the shut down of these stations for anything longer than a very brief period would be disastrous. Therefore, it is unlikely that all would be shut down, and this may be the reason little has been said about this issue beyond the initial statement.

B. LONG TERM EFFECTS

In the longer run, the accident is likely to affect the Soviet energy system in several ways. Given the seriousness of the accident, it will inevitably slow down the Soviet nuclear power program, at least over the next five years. However, it is very unlikely that it will be halted because of the government's strong commitment to nuclear power. The slowdown can be projected from several factors:

1) Questions about the safety of nuclear power will begin to be raised openly for the first time, as will questions concerning the RMBK design. The necessary investigations, etc. will require time.

2) In all likelihood, reinforced concrete containment domes will be required, perhaps not at all stations, but this would divert resources from new reactor and station construction.

3) Some plans for nuclear heating plants and other plants near large population centers may be altered, adding further delays.

These delays in the nuclear power program will lead to greater use of conventional thermal power plants, which will require increased shipments of fossil fuels from Siberia. Also, if the accident affects nuclear power development in Eastern and Western Europe, this will increase the demand for Soviet fossil fuel exports, particularly gas, there.

It is also likely that the accident at Chernobyl' will strengthen the position of proponents of ultra-high-voltage lines to move power from large mine-mouth coal-fired stations in the East to supply electricity needs in the European area. A debate has existed for some years on the desirability of a 1500 KVA direct current line between Ekibastuz in Kazakhstan and the Center. Although its construction has been included in the last couple of five-year plans, little progress has been made. After its construction was included in the most recent draft 12th Five-Year Plan, opponents, principally in Gosstroy, the State Construction Agency, argued that the situation has fundamentally changed since the project was first proposed; because of the development of nuclear power in the European USSR, the project is no longer needed and should be dropped (Sotsialiticheskaya industriya, January 12, 1986, p. 2). The analysis in Chapter VII also indicates this to be the case.

Proponents then countered with the argument that the line was necessary, not only for supplying power to the energy-short European USSR as originally proposed, but for maneuverability, to transfer power in both directions, making fuller use of the nuclear power stations (Pravda, February 5, 1986, p. 2). The safety of nuclear power, although existing as an issue, could not be debated openly because of the commitment of the government to nuclear power. The Chernobyl' accident may bring this issue to the surface and slowdown the pace of nuclear development, and therefore strengthen the position of the proponents of ultra-high-voltage lines.

The severity of the slowdown in nuclear power development is difficult to gauge this early. The new 12th Five-Year Plan called for nuclear power production to reach 390 billion KWH in 1990, or some 21 percent of total production (Sotsialiticheskaya industriya, March 9, 1986, p. 3), up from 170 billion KWH in 1985. It is yet too early to discuss the impact of slowdown. But for every 100 billion KWH they don't produce through nuclear generation, they will require the shipment from Siberia of an extra 10 million tons of fossil fuels, or 8 BCM of natural gas, or 7 million tons of crude oil, or 12 million tons of Ekibastuz coal.

References

Alisov, N.V. 1976. "Spatial Aspects of the New Soviet Strategy of Intensification of Industrial Production," Soviet Geography: Review and Translation, vol. 20, no.1 (January 1979), pp.1-6.

Atlas zheleznykh dorog. 1984. Moscow: Glavnoye upravleniye geodezii i kartografii pri Sovete ministrov SSSR.

Biryukov, V.Ye. et al. 1982. Ratsionalizatsiya perevozok gruzov, Moscow: Znaniye.

Biryukov, V.Ye. 1981. Transport v odinnatsatoy pyatiletke, Moscow: Znaniye.

Brents, A.D. et al. 1985. "Analiz pribyli i rentyabel'nosti v magistral'nom transporte gaza," Ekonomika i upravleniye v gazovoy promyshlennosti, no.3.

Campbell, Robert W. 1980. Soviet Energy Technology: Planning, Policy, Research and Development, Bloomington: Indiana University Press.

Campbell, Robert W. 1983. "Energy," in Abram Bergson and Herbert S. Levine (eds.), The Soviet Economy: Toward the Year 2000, London: Allen and Unwin, pp.191-217.

Campbell, Robert W. 1976. Trends in the Soviet Oil and Gas Industry, Baltimore: Johns Hopkins University Press.

CIA (Central Intelligence Agency). 1978. USSR: Development of the Gas Industry, ER 78-10393 (July).

CIA (Central Intelligence Agency). 1980. USSR: Coal Industry Problems and Prospects, ER 80-10154 (February).

CIA (Central Intelligence Agency). 1982. "Sluggish Soviet Steel Industry Holds Down Economic Growth," in U.S. Congress, Joint Economic Committee, Soviet Economy in the 1980s: Problems and Prospect, Washington, D.C.: U.S. Government Printing Office, pp.195-215.

CIA (Central Intelligence Agency). 1985. USSR Energy Atlas.

Danilov, A.D. et al. 1983. Ekonomicheskaya geografiya SSSR, Moscow: Vysshaya shkola.

Danilov, S.K. 1977. Ekonomicheskaya geografiya transporta SSSR, Moscow: Transport.

Davies, R.W. 1974. "Economic Planning in the USSR," in Morris Bornstein (ed.), Comparative Economic Systems: Models and Cases, 3rd ed., Homewood, Ill.: Richard D. Irwin, Inc., pp.266-290.

Davydov, B.N., and L.T. Artyukhova. 1979. "Ob urovne i territorial'noy differentsiatsii optovykh promyshlennykh tsen na topochnyy mazut," Khimiya i tekhnologiya topliv i masel, no. 7, pp.24-27.

Dienes, Leslie. 1983a. "Soviet Energy Policy and the Fossil Fuels," in Robert G. Jensen, Theodore Shabad, and Arthur W. Wright (eds.), Soviet Natural Resources in the World Economy, Chicago: University of Chicago Press, pp.275-295.

Dienes, Leslie. 1983b. "The Regional Dimension of Soviet Energy Policy," in Robert G. Jensen, Theodore Shabad, and Arthur W. Wright (eds.), Soviet Natural Resources in the World Economy, Chicago: University of Chicago Press, pp.385-410.

Dienes, Leslie, and Theodore Shabad. 1979. The Soviet Energy System: Resource Use and Policies, Washington, D.C.: Winston and Sons.

Dohan, Michael R. 1979. "Export Specialization and Import Dependence in the Soviet Economy, 1970-77," in U.S. Congress, Joint Economic Committee, _Soviet Economy in a Time of Change_, vol. 2, Washington, D.C.: U.S. Government Printing Office, pp.342-395.

Dubinskiy, P.K. 1977. _Analiz raboty zheleznodorozhnogo transorta ugol'noy promyshlennosti v devyatey pyatiletke_, Moscow: TsNIEIugol'.

Dubinskiy, V.G. 1977. _Ekonomike razvitiya i razmeshcheniya nefteprovodnogo transporta v SSSR_, Moscow: Nedra.

Ekonomicheskaya gazeta. 1981. No. 46, November.

Ekonomicheskaya gazeta. 1984. No. 19, May.

Ekonomicheskaya gazeta. 1984. No. 49, December.

Ekonomicheskaya gazeta. 1985. Nos. 1, 5, January.

Ekonomicheskaya gazeta. 1986. No. 3, January.

Energy Information Administration. 1983. _Petroleum Supply Annual_, Washington, D.C.: U.S. Department of Energy.

Fedorov, N.A. _et al_. 1981. "Osnovnoyye printsipy gazosnabzheniye narodnogo khozyaystva," _Gazovaya promyshlennost'_, no. 12, p.7.

Feshbach, Murray, and Stephen Rapawy. 1976. "Soviet Population and Manpower Trends and Policies," in U.S. Congress, Joint Economic Committee, _Soviet Economy in a New Perspective_, Washington, D.C.: U.S. Government Printing Office, pp.113-154.

Foreign Broadcast Information Service. 1984. _FBIS Daily USSR Report_, 8 August, p.4.

Furman, I.Ya. 1978. _Ekonomika magistral'nogo transporta gaza_, Moscow: Nedra.

Gankin, M.Kh. _et al_. 1972. _Perevozki gruzov (tendensey razvitiya i putiratsionalizatsiy)_, Moscow: Transport.

Green, Milford B., and Ronald L. Mitchelson. 1981. "Spatial Perspective of the Flows through the Southeast Electrical Transmission Network," The Professional Geographer, vol. 33, no. 1 (February), pp.83-94.

Green, Milford B., and Matthew J. Sagers. 1985a. "Changes in Soviet Natural Gas Flows: 1970-1985," The Professional Geographer, vol. 37, no. 3.

Greenslade, Rush V. 1976. "The Real Gross National Product of the USSR, 1950-1975," in U.S. Congress, Joint Economic Committee, Soviet Economy in a New Perspective, Washington, D.C.: U.S. Government Printing Office, pp.270-300.

Grigor'yev, A.N. et al. 1984. "Optimizatsiya perevozok nefteproduktov", Zheleznodorozhniy transport, no. 2, pp.60-64.

Grigor'yev, V.A., and V.M. Zorin (eds.). 1980. Teploenergetika i teplotekhnika, Moscow: Energiya.

Gustafson, Thane. 1982. "Soviet Energy Policy," in U.S. Congress, Joint Economic Committee, Soviet Economy in the 1980s: Problems and Prospects, Part I, Washington, D.C.: U.S. Government Printing Office, pp.431-456.

Gustafson, Thane. 1985. "The Origins of the Soviet Oil Crisis, 1970-1985," Soviet Economy, vol. 1, no. 2 (April-June), pp.103-135.

Hardt, John, et al. 1966. "Institutional Stagnation and Changing Economic Strategy in the Soviet Union," in Joint Economic Committee, U.S. Congress, New Directions in the Soviet Economy, Part I, Washington, D.C.: U.S. Government Printing Office, pp.19-62.

Hewett, E.A. 1984. Energy, Economics, and Foreign Policy in the Soviet Union, Washington, D.C.: The Brookings Institution.

Houston, Cecil. 1969. "Market Potential and Potential Transportation Costs: An Evaluation of the Concepts and their Surface Patterns in the USSR," _The Canadian Geographer_, vol. 13, no. 3, pp.216-236.

Hunter, Holland. 1968. _Soviet Transport Experience: Its Lessons for Other Countries_, Washington, D.C.: The Brookings Institution.

Hunter, Holland, and Deborah A. Kaple. 1982. "Transport in Trouble," in Joint Economic Committee, U.S. Congress, _Soviet Economy in the 1980s: Problems and Prospects_, Part I, Washington, D.C.: U.S. Government Printing Office, pp.216-241.

Hunter, Holland, and Deborah A. Kaple. 1983. _The Soviet Railroad Situation_, Washington, D.C.: Wharton Econometrics Forecasting Associates.

Huzinec, George. 1978. "The Impact of Industrial Decision-Making upon the Soviet Urban Hierarchy," _Urban Studies_, vol. 15, no. 2 (June), pp.139-148.

Izvestiya, 4 May 1984.

Izvestiya, 6 July 1984.

JPRS (Joint Publications Research Service). 1982. "Map of USSR Electric Power Stations," JPRS L/10822, 22 September.

JPRS (Joint Publications Research Service). 1984. _Soviet Union: Transportation_, JPRS UTR-84-023, 7 August, p.32.

Kal'chenko, V.N. _et al._ 1974. _Ekonomika gazovoy promyshlennost'_, Kiev: Tekhnika.

Karchik, V.G. 1985. "Optimizatsiya planov perevozok nefteproduktov," _Zheleznodorozhniy transport_, no. 11 (1985), pp.62-66.

Kazakhstanskaya Pravda, 25 December 1983.

Kazanskiy, N.N. (ed.). 1980. Geografiya putey soobshcheniya, Moscow: Transport.

Kibalchich, O.A. 1983. "Comments," Soviet Geography: Review and Translation, vol. 24, no. 10 (December), pp.724-727.

Kirichenko, V.N. 1981. "Odinnatstataya pyatiletka: strukturnyye sdvigi v ekonomike," Izvestiya Akademii Nauk SSSR: Seriya Ekonomicheskaya, no. 2, pp.5-6, 13.

Kontorovich, Vladimir. 1982. A Case Study of Transportation in the Urals, West Siberia and North Kazakhstan, Washington, D.C.: Wharton Econometric Forecasting Associates, December.

Kurtzweg, Laurie R., and Albina Tretyakova. 1982. "Soviet Energy Consumption: Structure and Future Prospects," in U.S. Congress, Joint Economic Committee, Soviet Economy in the 1980s: Problems and Prospects, Part I, Washington, D.C.: U.S. Government Printing Office, pp.355-390.

Lydolph, Paul E. 1979. Geography of the USSR: A Topical Analysis, Elkhart Lake, Wisconsin: Misty Valley.

Maslov, V.O., and M.D. Perova. 1976. "Tselesoobraznost' ispol'zovaniya razlichnykh vidov transporta dlya perevozki svetlykh nefteproduktov," Transport i khraneniye nefti i nefteproduktov, no. 10, pp.30-35.

Meyerhoff, Arthur A. 1983. "Soviet Petroleum: History, Technology, Geology, Reserves, Potential and Policy," in Robert G. Jensen, Theodore Shabad, and Arthur W. Wright (eds.), Soviet Natural Resources in the World Economy, Chicago: University of Chicago Press, pp.306-362.

Minieka, Edward. 1978. Optimization Algorithms for Networks and Graphs, New York: Marcel Dekker.

Mints, A.A. 1976. "A Predictive Hypothesis of Economic Development in the European Part of the USSR," Soviet Geography: Review and Translation, vol. 17, no.1 (January), pp.1-28.

Mitrofanov, V.F. 1977. "Razvitiye gruzovykh peravozok v desyatoy pyatiletke," Zheleznodorozhnyy transport, no. 6, pp.81-88.

Mozhin, V. 1983. "Ratsional'noye razmeshcheniye proizvoditel'nykh sil i sovershenstvovaniye territorial'nykh proportsiy", Planovoye khozyaystvo, no. 4, pp.3-12.

Narkhoz SSSR .... TsSU (Tsentral'noye statisticheskoye upravleniye priSovete Ministrov SSSR). Narodnoye khozyaystvo SSSR v ... godu, statisticheskiy yezhegodnik, Moscow: Finansy i statistika: ... (annual); cited as Narkhoz SSSR with appropriate year.

Neftyanoye khozyaystvo. 1977. No. 10.

Neftyanoye khozyaystvo. 1985. No. 5.

Nekrasov, A.M. and A.A. Troitskiy. 1981. Energetika SSSR v 1981-1985 godakh, Moscow: Enerogizdat.

Neporozhniy, P.S. 1980. 60 let Leninskogo plana GOELRO, Moscow: Energiya.

"News Notes," SG-....: Theodore Shabad. "News Notes," Soviet Geography: Review and Translation:

1969: vol. 10, no. 7 (September), p.423.
1975: vol. 16, no. 1 (February), p.122.
1976: vol. 17, no. 10 (December).
1977: vol. 18, no. 9 (November), p.702.
1978: vol. 19, no. 4 (April), p.275.
1979: vol. 20, no. 2 (February), p.126.
1979: vol. 20, no. 5 (May), p.332.
1980: vol. 21, no. 3 (March), pp.187-190.
1981: vol. 22, no. 4 (April), p.275.
1982: vol. 23, no. 3 (March), p.197.
1982: vol. 23, no. 4 (April), pp.177-283, 287-291, 282-287, 278.
1982: vol. 23, no. 5 (May), p.374.
1982: vol. 23, no. 7 (September), pp.540-548.
1982: vol. 23, no. 10 (December), pp.769-772.
1983: vol. 24, no. 3 (March), pp.249-256.

1983: vol. 24, no. 4 (April), pp.216, 318.
1983: vol. 24, no. 8 (October), pp.625-627.
1983: vol. 24, no. 9 (November), pp.703-707.
1983: vol. 24, no. 10 (December), pp.775-777.
1984: vol. 25, no. 4 (April), pp.264-268, 272, 275, 277, 280-284, 290-291.
1984: vol. 25, no. 7 (September), p.547.
1985: vol. 26, no. 4 (April), pp.288-294, 294-300, 303.
1986: vol. 27, no. 4 (April),

Office of Technology Assessment, U.S. Congress. 1981. Technology and Soviet Energy Availability, Washington, D.C.: U.S. Government Printing Office.

Oil and Gas Journal. 1984. Vol. 81, no. 1, pp.25-28.

Osipov, N. 1985. "Perspektivy eksporta SSSR," Vneshnaya torgovlya, no. 2, pp.3-9.

Perova, M.D. and V.O. Maslov. 1984. "Razvitiye nefteprodukto-provodnogo transportov SSSR i SShA", Transport i khraneniye nefteproduktov i uglevodorodnogo syr'ya, vol. 5, pp.7-10.

PlanEcon Data Bank, 1986, p. 5-20.

PlanEcon Report, 1986, nos. 19-20.

Pravda, 30 June 1983.

Pravda, 5 February 1986.

Pyzhkov, N. 1982. "Nekotoryye voprosy planovogo rukovodstva ekonomikoy," Planovoye khozyaystvo, no. 8 (August), pp.3-4.

Rodgers, Allan L. 1983. "Commodity Flows, Resource Potential and Regional Economic Development: The Example of the Soviet Far East," in Robert G. Jensen, Theodore Shabad and Arthur W. Wright (eds.), Soviet Natural Resources in the World Economy, Chicago: University of Chicago Press, pp.188-213.

Rumer, Boris and Stephen Sternheimer. 1982. "The Soviet Economy: Going to Siberia?" Harvard Business Review, no. 82111 (January-February), pp.1-10.

Runova, T.G. 1976. "The Location of the Natural Resource Potential of the USSR in Relation to the Geography of Productive Forces," Soviet Geography: Review and Translation, vol. 17, no. 2 (February), pp.73-85.

Ryl'skiy, V.A. 1981. Regional'niye problemy razvitiya energetika i elektrifikatsii SSSR, Moscow: Ekonomika.

Ryps, G.S. 1978. Ekonomicheskiye problemy raspredeleniya gaza, Leningrad: Nedra.

Sagers, Matthew J. 1984. Refinery Throughput in the USSR, Center for International Research, CIR Staff Paper No. 2, Washington, D.C.: U.S. Bureau of the Census.

Sagers, Matthew J., and Milford B. Green. 1982. "Spatial Efficiency in Soviet Electrical Transmission," Geographical Review, vol. 72, no. 3 (July), pp.291-303.

Sagers, Matthew J., and Milford B. Green. 1984. "Coal Movements in the USSR," Soviet Geography, vol. 25, no. 10 (December), pp.713-733.

Sagers, Matthew J., and Milford B. Green. 1985a. "Transport Constraints in Soviet Petroleum," Energy Policy, vol. 13, no. 4 (August), pp.371-380.

Sagers, Matthew J., and Milford B. Green. 1985b. "The Freight Rate Structure on Soviet Railroads," Economic Geography, vol. 61, no. 4, pp.305-322.

Sagers, Matthew J., and Albina Tretyakova. 1985. Restructuring the Soviet Petroleum Refining Industry, CIR Staff Paper No. 4, Washington, D.C.: Bureau of the Census.

Schroeder, Gertrude E. 1985. "The Slowdown in Soviet Industry, 1976-1982," Soviet Economy, vol. 1, no. 1, pp.42-74.

Sedykh, A.D., and B.A. Kuchin. 1983. Upravleniye nauchno-tekhnicheskim progressom v gazovoy promyshlennosti, Moscow: Nedra.

SEV, ... Statisticheskiy yezhegodnik stran-chlenov soveta ekonomicheskoyvzaimopomoshchni, Moscow: Finansy i statistika, ... (annual); (volumes for 1976 and 1981 were used).

Shafirkin, B.I. 1978. Ekonomicheskiy spravochnik zheleznodorozhnika, chast II, Moscow: Transport.

Shcherbina, B.Ye. et al. 1981. Otechestvennyy truboprovodnyy transport, Moscow, Nedra.

Shul'ga, A.M. 1974. Sebestoimost' zheleznodorozhnykh perevozok i puti yeye snizheniya, Moscow: Transport.

Sidorenko, M.V., Ye.I. Kolosov, and I.Ya. Furman. 1983. "Regional'nyye i mezhotraslvyye problemy osvoyeniya gazovykh mestorozhdeniy i razvitiya sistem dal'nego transporta gaza," Ekonomika, organizatsiya i upravleniye v gazovoy promyshlennosti, no. 7.

Sotsialiticheskaya industriya, January 3, 1985.

Sotsialiticheskaya industriya, October 7, 1985.

Sotsialiticheskaya industriya, January 12, 1986.

Sovetskaya Rossiya, June 16, 1985.

Sredin, V.V., and G.A. Lastovkin. 1972. "Ekonomicheskiye pokazateli kombinirovannoy ustanovki LK-6U", Khimiya i technologiya topliv i masel, vol. 12, pp.30-33.

Stern, Jonathan P. 1981. "Western Forecasts of Soviet and East European Energy over the Next Two Decades (1980-2000)," in U.S. Congress, Joint Economic Committee, Energy in Soviet Policy, Washington, D.C.: U.S. Government Printing Office, pp.55-79.

Stern, Jonathan P. 1983. "Soviet Natural Gas in the World Economy," in Robert G. Jensen, Theodore Shabad and Arthur W. Wright (eds.), Soviet Natural Resources in the World Economy, Chicago: University of Chicago Press, pp.363-384.

Stroitel'tsvo truboprovodov, 1984. no. 1, p.4; no. 2, pp.2-4.

Styrikovich, M.A., and S.Ya. Chernyavskiy. 1979. "Puti razvitiya i rol' yadernoy energetiki v perspektivnom energobalanse mira i ego osnovnykh regionov," Joint U.S.-Soviet Seminar under Energy Agreement.

Summary of World Broadcasts, 16 August 1985.

Summary of World Broadcasts, 28 February 1986.

Teploenergetika, 1986. no. 2, p. 3.

Tretyakova, Albina. 1980. "Coal Industry," unpublished working paper. Center for International Research, Washington, D.C.: U.S. Bureau of the Census.

Tretyakova, Albina. 1985. Expenditures on Fuels in the USSR to 1990, Foreign Economic Report, No. ..., Washington, D.C.: Bureau of the Census.

Ugol'. no. 3 (1986).

Ugol'. no. 7 (1985), pp.23-24.

United National Statistical Office. 1983. 1981 Yearbook of World Energy Statistics, New York: United Nations.

Venikov, V. 1982. Transport energiy, Moscow: Znaniye.

Vestnik akademii nauk SSSR. 1985. No. 12.

Vlasov, A.V. 1984. "Bor'ba s poteryami nefteproduktov pri transportirovanii i khranenii", _Transport i khraneniye nefteproduktov i uglevodorodnogo syr'ya, tematicheskiy obzor_, Moscow: TsNIII i TEINNP.

Vneshnaya torgovlya SSSR za .... god, statisticheskiy obzor, Ministerstvo vneshney torgovli SSSR. Moscow: Mezhdunarodnoyye otnosheniya, ... (annual).

Voloboy, P.V., and V.A. Popovkin. 1972. _Problemy territorial'noy spetsializatsii i kompleksnogo rozvitku narodnogo gospodarstva Ukrainskoy RSR_, Kiev: Naukova dumka.

WEFA, Wharton Econometric Forecasting Associates, Inc. 1983. _Data Bank_. (Now maintained by Plan Econ, Inc.)

Wilson, David. 1983. _The Demand for Energy in the USSR_, Totowa, New Jersey: Rowman and Allenheld.

Yufina, V.A. 1978. _Truboprovodnyy transport nefti i gaza_, Moscow, Nedra.

ZumBrunnen, Craig, and Jeffrey Osleeb. 1986. _The Soviet Iron and Steel Industry_, Totowa, New Jersey: Rowman and Allenheld.

Index

TERMS

AUTHORS

PLACES